Sustainable Development Goals Series

The **Sustainable Development Goals Series** is Springer Nature's inaugural cross-imprint book series that addresses and supports the United Nations' seventeen Sustainable Development Goals. The series fosters comprehensive research focused on these global targets and endeavours to address some of society's greatest grand challenges. The SDGs are inherently multidisciplinary, and they bring people working across different fields together and working towards a common goal. In this spirit, the Sustainable Development Goals series is the first at Springer Nature to publish books under both the Springer and Palgrave Macmillan imprints, bringing the strengths of our imprints together.

The Sustainable Development Goals Series is organized into eighteen subseries: one subseries based around each of the seventeen respective Sustainable Development Goals, and an eighteenth subseries, "Connecting the Goals", which serves as a home for volumes addressing multiple goals or studying the SDGs as a whole. Each subseries is guided by an expert Subseries Advisor with years or decades of experience studying and addressing core components of their respective Goal.

The SDG Series has a remit as broad as the SDGs themselves, and contributions are welcome from scientists, academics, policymakers, and researchers working in fields related to any of the seventeen goals. If you are interested in contributing a monograph or curated volume to the series, please contact the Publishers: Zachary Romano [Springer; zachary.romano@springer.com] and Rachael Ballard [Palgrave Macmillan; rachael.ballard@palgrave.com].

Pilar Pérez-Hernandez ·
Rocío Huerta Cuervo
Editors

Social Well-Being and the challenges in the Sustainable Development Goals

Perspectives from Mexico

Springer

Editors
Pilar Pérez-Hernandez
Instituto Politécnico Nacional
Mexico City, Mexico

Rocío Huerta Cuervo
National Polytechnic Institute
Mexico City, Mexico

ISSN 2523-3084 ISSN 2523-3092 (electronic)
Sustainable Development Goals Series
ISBN 978-3-032-14293-1 ISBN 978-3-032-14294-8 (eBook)
https://doi.org/10.1007/978-3-032-14294-8

This Springer imprint is published by the registered company Springer Nature Switzerland AG
The registered company address is: Gewerbestrasse 11, 6330 Cham, Switzerland

Introduction

The adoption of the **Sustainable Development Goals (SDGs)** at the 70th United Nations General Assembly in 2015 did not represent a mere change in acronyms from the Millennium Development Goals (MDGs); it signified a profound transformation in the global development roadmap. While the MDGs focused on unidimensional metrics that, although effective in mobilizing resources, tended to simplify complex structural problems, the SDGs emerge as a **universal and systemic call** to eradicate poverty, protect the planetary ecosystem, and guarantee peace and prosperity by the year 2030.

The SDGs emphasize the need to advance the well-being of people from the places where they live, expand the methodologies to evaluate progress, and seek to incorporate local actors in the generation of diagnoses, strategies, and policies to address public problems.

This work, coordinated by **Pilar Pérez-Hernández** and **Rocío Huerta Cuervo**, presents a critical and transdisciplinary review of Mexico's progress toward this global commitment. The content and focus of each chapter are detailed below:

Sustainable Development and Institutional Capacity: Monitoring the SDGs in Mexico. María del Pilar Monserrat Pérez Hernandez y Rocio Huerta Cuervo analyze the evolution of international efforts toward sustainable development and social rights. The text emphasizes that achieving the SDGs depends on strong institutions, reliable data, local participation, and political commitment, particularly at the municipal level.

Municipal Institutional Capacities and the SDGs: Rocío Huerta Cuervo, Evelia Alarcón Rojas, and Luis Pablo Murillo Tovar analyze how the construction of well-being is materialized at the local level. They argue that lags in municipalities condition the fulfillment of the SDGs, noting that of 2,097 municipalities studied, 1,425 possess low or very low capacities.

SDGs, Planetary Citizenship, and Humanist Education: Conde Gaxiola, Rivas Tovar, and Rivera González explore gaps in the educational system. At the center of capacity building is education as the intangible support for human performance across all sectors.

Beyond the SDGs: Degrowth as an Alternative: Authors Rivera, Rivera, Carrillo, and Cadena López present a critical discussion on the development model, proposing degrowth as a path forward against unsustainable consumption patterns.

Green Finance Status in Major Latin American Economies: Escamilla-García and Morales Castro offer a critical review of the financial mechanisms necessary to support the transition to a sustainable economy in the region.

Sustainability of Communal Enterprises in the New Institutional Economy: Maldonado, Téllez-Cabrera, and Hernández-Mendoza examine how communal organizational forms in Mexico represent models of resilience and sustainable management.

Governance in the Health Sector and the SDGs: Pérez-Hernández, Castillo, and Gómez analyze SDG 3 in light of the COVID-19 crisis, which halted and reversed progress, causing millions of deaths and revealing failures in health management. They identify structural gaps in indicators and weak efforts toward meeting "Good Health and Well-being" targets.

Grocery Shopping as a Means to Responsibly Empower Women: Cruz Reyes, Rojas Alarcón, and Castillo Malagón link daily life and household economics with SDG 5, seeking strategies to reduce gender inequality.

Social Entrepreneurship in Tourism: Tejeida Padilla, Pérez Matamoros, Patiño Abuela, and Badillo Piña explore tourism through a sustainability lens and analyze the critical relationship between tourism and water resources, where centralized legislation often ignores local realities.

Integrated Causal Map of MSMEs in the Creative Sector: Macías-Barreto, Aguilar-Fernández, Patiño-Ortiz, and Álvarez-Cedillo propose tools to promote these economic units in Mexico.

Decent Work: A Transdisciplinary SDG: The same team (Macías-Barreto et al.) addresses SDG 8, analyzing the implications of full and productive employment against a reality where informality remains a major challenge.

A fundamental contribution of this analysis is the evaluation of Mexico's national SDG platform. Severe limitations are detected: the platform presents only **11 indicators with a municipal breakdown**, despite institutions like INEGI possessing much broader databases. Furthermore, there is a gap between CONEVAL's information on multidimensional poverty and the goals defined on the platform, hindering a true assessment of social welfare.

Conclusion and Future Perspectives

Mexico faces a complex scenario. Advancing toward the 2030 Agenda demands commitment among political forces, business owners, and citizens. However, conditions such as a **weak rule of law**, generalized violence, and political confrontation hinder this path.

The work concludes that to advance the SDGs, it is imperative to strengthen the role of local actors. Through these studies, the **Instituto Politécnico Nacional** seeks to offer a critical vision to interpret and transform reality toward more sustainable and democratic societies. This work is a call to action to **“empower the local”**. The SDGs will only be achievable in a democratic society where respect for human rights and substantive freedoms are the norm.

Pilar Pérez-Hernandez
Rocío Huerta Cuervo

Contents

1 Sustainable Development and Institutional Capacity: Monitoring the SDGs in Mexico

Pilar Pérez-Hernandez and Rocío Huerta Cuervo

The new global development roadmap, adopted at the 70th United Nations General Assembly in September 2015 in New York (United States), has been named the Sustainable Development Goals (SDGs). Since then, it has become the fundamental reference point for global development through 2030. It was made clear that changing the world requires urgent cooperation and participation among governments, businesses, international organizations, world leaders, academics, and citizens (Gómez 2017).

The SDGs are a universal call to end poverty, protect the planet, and ensure that by 2030 all people enjoy peace and prosperity. The SDGs did not emerge overnight; they are part of a long series of initiatives launched in the 1990s by governments, the United Nations, and international financial institutions aimed at reducing global poverty. One of the earliest was the "Earth Summit," held from June 3 to 14, 1992, in Rio de Janeiro (Brazil), which brought together political leaders, diplomats, scientists, media representatives, and non-governmental organizations from 179 countries to focus on the impact of socioeconomic activities on the environment. The main objective was to produce a broad agenda on environmental and development issues to help guide international cooperation and development policy in the 21st century (UN 2000).

P. Pérez-Hernandez (✉) · R. Huerta Cuervo
IPN—Instituto Politécnico Nacional, Mexico City, Mexico
e-mail: mpperez@ipn.mx

R. Huerta Cuervo
e-mail: rhuerta@ipn.mx

At the Earth Summit, it was concluded that sustainable development was achievable for all people worldwide, whether at the local, regional, national, or international level, but a new perspective and decision—making on how production, consumption and works take place is essential. Among its outcomes was the implementation of Agenda 21, which included recommendations for new educational methods, new ways to preserve natural resources, and new approaches to participating in a sustainable economy (UN 2000).

The second was the International Conference on Women held in Beijing (China) in September 1995, which marked a turning point for the global agenda on gender equality. Representatives from 189 countries unanimously established a program to empower women in 12 critical areas: women and poverty; education and training for women; women and health; violence against women; women and armed conflict; women and the economy; women in power and decision-making; institutional mechanisms for the advancement of women; women's human rights; women and the media; women and the environment; and the protection of girls' rights (UN 1996).

The third was the International Conference on Population and Development held in September 1994 in Cairo (Egypt), where 179 states participated to finalize a program of action on population and development for the next 20 years. The objectives included education, especially for

P. Pérez-Hernandez and R. Huerta Cuervo (eds.), *Social Well-Being and the Challenges in the Sustainable Development Goals*, Sustainable Development Goals Series,
https://doi.org/10.1007/978-3-032-14294-8_12

girls, and achieving greater reductions in infant and maternal mortality rates. The conference also addressed issues related to population, the environment, and consumption patterns; the family; internal and international migration; the prevention and fight against the HIV/AIDS pandemic; information, education, and communication; technology; research; and development (UN 1995).

These conferences enabled governments to address issues related to social rights, sustainable development, and environmental protection in their agendas. The result of these initiatives was the establishment of the political will necessary to seek coordinated solutions between developed and developing countries. The most significant achievement was addressing poverty, understood as an obstacle to development that could only be overcome within the framework of so-called North-South cooperation (UN 1995).

In September 2000, representatives from 189 countries met at the so-called Millennium Summit, which gave rise to the Millennium Development Goals (MDGs). Eight goals were established: eradicate poverty and hunger; achieve universal primary education; promote gender equality and empower women; reduce child mortality; improve maternal health; combat HIV infection, malaria, and other diseases; ensure environmental sustainability; and foster a global partnership for development. These goals were key to establishing measurement indicators, which enabled feedback on the performance of policies, strategies, and implementation, allowing for improvements.

Importantly, the MDGs provided lessons for designing the criteria used in constructing the indicators for the SDGs approved in 2015. They encouraged more critical reflection on their management, that is, on governance for sustainable development at the local level (Fukuda-Parr 2016). The SDGs are the result of global commitments that require governments, private organizations, higher education institutions, citizens, and their organizations to work together to eliminate the causes of poverty and environmental degradation, creating conditions for the well-being and development of people worldwide (UN 2015; Alleyne et al. 2015; Birendra et al. 2021).

The transition from the Millennium Development Goals (MDGs) to the SDGs involved a critical reflection on measurement systems for tracking progress toward committed goals and objectives. While the MDGs "helped mobilize resources and political will... they reduced complex issues to one-dimensional measurement questions" (Hansson et al. 2019). The evaluation systems of various governments emphasized formal processes, neglecting comprehensive assessments of real improvements in people's well-being in their living environments. As noted by (Kemp et al. 2005; Loorbach 2010, as cited in Hansson 2019, p. 218), for SDG indicators to contribute to sustainability, it is necessary to consider the relationship between the scorecard (indicator results) and management functions at the local level, since together they must be able to "manage complexity rather than reduce it, contribute to integration rather than fragmentation, and create incentives for innovation to address future challenges rather than seek low-hanging fruit."

The SDGs propose systemic responses to a global and interconnected vision of sustainable development that addresses critical issues such as inequality and extreme poverty, unsustainable consumption patterns and environmental degradation, the strengthening of institutional capacities, as well as innovative processes of global solidarity that the MDGs overlooked. The SDGs propose systemic responses to a global and interconnected vision of sustainable development that addresses critical issues such as inequality and extreme poverty, unsustainable consumption patterns and environmental degradation, the strengthening of institutional capacities, as well as innovative processes of global solidarity that the MDGs overlooked (Gómez 2017).

Institutional capacities are understood as the assets that societies possess, enabling them to advance toward their established objectives (UNDP 2009). Institutional capacities are built collectively and are reflected in the rules (formal and informal), infrastructure, social coexistence, and wealth that a community is able to produce, which in turn shape the participation of individuals within it. At the core of capacity building and development is invariably education, as the intangible support for individuals' performance in all areas of participation. Transitioning from the MDGs to the SDGs has brought to the forefront of public agencies' concerns the importance of fostering institutional capacity building at the local level, since the construction of well-being is realized locally.

The SDGs incorporate innovative elements in their design, among which we can highlight the integration of the three dimensions of sustainable development: economic, social, and environmental. They prioritize the fight against poverty and hunger, but with a strong emphasis on the defense of human rights, gender equality, and the empowerment of women, addressing the reduction of inequalities within each country and between different states as a prevailing element, alongside the elimination of unsustainable consumption patterns. In addition, they incorporate a vision of inclusive and sustainable economic growth, respectful of the health of the planet and its people. An innovative feature is their universal character, with a holistic and interconnected vision in which each goal overlaps and mutually reinforces the others, with simultaneous multilevel action at local, regional, national, and global levels, aiming to build enhanced global solidarity and integrating the major agreements reached at world summits (Gómez 2017).

The SDGs defined 17 goals and 169 targets. As part of the commitment of each signatory country, the need was established to implement a monitoring platform for SDG compliance. In the case of Mexico, in 2015 (Government of the Republic 2018a), the Technical Committee (CTEODS) was established to monitor the SDGs. By June 2018, all 32 states of the Republic had SDG Monitoring and Implementation Bodies (ISIODS), and 300 municipalities had joined these activities. Although the cited report (Government of the Republic 2018a) identified challenges to advancing SDG compliance, including "ensuring greater inter-institutional coordination, both vertical and horizontal; institutionalizing mechanisms for effective participation and co-creation of public policies; consolidating the ownership of the agenda by different sectors of society; expanding communication and dissemination strategies for the agenda; training stakeholders at all levels of government; ensuring monitoring and review mechanisms for actions and public policies that transcend administrative periods; and having long-term planning with intermediate objectives" (Government of the Republic 2018a), it is clear that all of this requires a political and social environment of commitment among political forces, business leaders, and citizens in order to be achieved. Conditions such as a weak rule of law, widespread violence in some regions, and ongoing confrontation among key stakeholders do not provide the best scenario for addressing these challenges. The structural lag in the institutional capacities of municipalities poses a significant constraint to progress and compliance.

This text is based on the assumption that the SDGs are achievable objectives in democratic societies, where human rights are respected and people enjoy substantial freedoms. It is clear that, given the diversity of structural conditions in different countries, some will be better positioned to advance in their achievement than others, which is the major challenge they represent for national and subnational governments and for the organizations in society committed to them.

Below is a brief description of the content of the targets associated with the SDGs in Mexico. The platform called "Sustainable Development Goals" (hereinafter national platform) is analyzed, where the federal government reports on progress in the indicators associated with the SDGs. The aim is to broadly outline the information that is institutionally available in Mexico to assess progress.

While the national platform represents an effort to systematize information, there is no clear identification of the conceptual content of each SDG. As a result, some indicators presented on the platform do not appear to be linked to the specific content of the goal in question. It is also important to note that the platform presents, interchangeably, both management indicators and outcome indicators. The former are a fundamental source for decision-makers in each area of responsibility, while the latter are essential for society, as they allow for the identification of progress toward the goals and the challenges that remain for their achievement. It would be essential to make this distinction in the presentation of information.

It should also be noted that not all information in Mexico related to the SDG targets is consolidated on this platform. There is a large amount of data held by INEGI, derived from national population censuses, municipal government censuses, and records of public spending at the municipal level, which have not been included in the national platform. The platform presents a breakdown of only 11 indicators at the municipal level (Government of Mexico 2023), which is perhaps its main limitation. At the local level, the set of indicators for each SDG target can take on a new dimension, as stakeholders involved with a specific reality are able to interpret, reinterpret, and integrate a more holistic understanding of the results. This is the major leap that must be made in terms of information, and in this process, it is essential to strengthen the role of local actors in achieving the SDGs and to empower the local level.

For SDG 1, titled "End poverty in all its forms everywhere," the national platform includes, in addition to the percentage of the population living in poverty, the number of adults with land tenure rights; the number of people killed or missing due to disasters per 100,000 inhabitants; the proportion of public spending allocated to social protection, education, and health; the percentage of children under one year of age with basic vaccination coverage; the percentage of the population aged 65 and over receiving a pension or retirement benefit; the percentage of employed people with children under six years old who have access to daycare services; the percentage of the population lacking access to services of social security (SDG 2024), without updated information being available for all of them. The national platform proposes 8 targets with 18 indicators for this goal.

CONEVAL (2019) has summarized the concept of poverty as a multidimensional construct that incorporates various variables, including per capita current income, average educational lag per household, access to health services, social security, basic housing services, access to nutritious and quality food, degree of social cohesion, and accessibility to paved roads (CONEVAL 2019). As is well known, globally, poverty is measured by the number of people living on less than $1.90 per day. In Mexico, the extreme poverty lines for rural and urban areas were set in December 2023 at 1743.71 pesos per month (102.3 dollars) for rural areas and 2269.57 pesos per month (133.12 dollars) for urban areas. The rural and urban poverty lines were set at 3226.61 (189.25 dollars) and 4461.97 pesos (261.70 dollars) per month, respectively (CONEVAL 2023). CONEVAL (2020) provides the most up-to-date and detailed information on poverty conditions in Mexico. Data are available on people living in poverty, moderate poverty, extreme poverty and those not vulnerable, all disaggregated by municipality (CONEVAL 2020), but all data are derived from calculations that come from the household income expenditure survey (ENIGH for its acronym in Spanish), not from a census. It also breaks down poverty by income, housing characteristics and the availability of basic services, food insecurity, educational attainment, coverage of social security and health services, and a summary count of the number of people experiencing one, two, or three basic deprivations.

As can be seen, there is a gap between the information presented by CONEVAL and the targets defined on the national platform. SDG 1 is the overarching goal that connects all the SDGs, and therefore, it would be important for the elements currently reported by CONEVAL to serve as the basis for its measurement and evaluation.

Regarding SDG 2, "End hunger, achieve food security and improved nutrition, and promote sustainable agriculture," the national platform includes 5 targets and 11 indicators for this goal. Among the most relevant are: proportion of the population experiencing food insecurity by sex, age, and region, which are already included in CONEVAL's poverty concept; children under 5 years with stunted growth; children under five with malnutrition (obesity and wasting); prevalence of anemia in pregnant women aged 15–49; and maize grain yield. In this SDG, in addition to the food poverty indicators already included by CONEVAL, central measurements are the percentage of children with malnutrition and pregnant women with anemia, but reported at the municipal level. The inclusion of maize grain yield as an indicator in this section is not entirely clear; rather, it would be essential to have municipal-level information on the percentage of hectares used for agriculture relative to the municipal total, and the percentage of those hectares engaged in sustainable production (with a clear definition of the concept).

For SDG 3, "Ensure healthy lives and promote well-being for all at all ages," there are 10 targets with 30 indicators, including: maternal mortality ratio, births attended by skilled health personnel, under-five mortality rate, neonatal mortality rate, number of new HIV infections per 1000 inhabitants by sex, age, and locality, HIV prevalence rate per 100,000 inhabitants, HIV prevalence among people aged 14–49, mortality rate from cardiovascular diseases, diabetes, cancer, and chronic respiratory diseases, deaths by suicide, women with satisfied demand for contraceptive methods aged 15–49, unmet needs for contraceptive methods among women aged 15–49, fertility rate among girls aged 10–14 per 1000 girls in that age group, fertility rate among adolescents aged 15–19, percentage of women aged 15–19 who are mothers, mortality rate attributable to unsafe water and poor sanitation, mortality rates attributable to unintentional poisonings, density and distribution of health personnel, percentage of the population lacking access to health services, percentage of the population who can reach a hospital in less than two hours, percentage of the population who did not spend money on health services the last time they sought medical care, percentage of households with catastrophic health expenditures, percentage of the population who are overweight and obese by sex, age, and state. All of these are highly relevant health indicators, but their data, in full detail are not publicly available in an updated and municipal-referenced manner. In some cases, it is necessary to submit information requests to state public information access agencies or the Ministry of Health to obtain them. The platform provides national or state-level data, but from previous years.

In Mexico, the leading cause of death among people over 45 years old is heart disease, followed by diabetes mellitus. Both diseases are associated with unhealthy lifestyles, such as physical inactivity and poor nutrition (INEGI 2023). The third leading cause of death is malignant tumors. The main cause of death among children (ages 5–14) is accidents, while among people aged 15–44, it is homicide, which is linked to the increase in violence throughout the country. Progress on the SDGs requires targeted attention according to the specific municipality and region.

For SDG 4, "Ensure inclusive and equitable quality education and promote lifelong learning opportunities for all," the national platform includes 9 targets with 27 indicators. Beyond indicators on literacy, coverage at each educational level, and completion rates, emphasis is placed on the percentage of schools with basic facilities such as piped water, handwashing sinks, ramps for people with disabilities, internet access, and computer equipment. Regarding the results of knowledge and competency assessments, no includes no indicator, nor does it address the suitability of teachers and school administrators. In 2019 (DOF 2019), the Constitution was amended to eliminate the evaluation processes for teachers and administrators, leaving two factors associated with educational quality without any data. For this SDG, information is provided on five indicators with breakdowns at the municipal level. It is important to reiterate the need for indicators on academic achievement and the suitability of teaching and administrative staff at the municipal level.

SDG 5, titled "Achieve gender equality and empower all women and girls," includes 10 targets and 30 indicators, none of which are presented with municipal-level breakdowns. This SDG emphasizes crimes against women, such as physical, psychological, and sexual assaults, as well as the number of hours women work at home without pay, whether caring for household members or performing domestic tasks. It also reports indicators such as the number of hours women work for pay, the proportion of women in leadership positions, and the wage gap between women and men according to their educational level. Several indicators relate to women's participation in representative bodies at various levels of government, as well as the proportion of women with property rights over agricultural land.

The reality for women, especially in municipalities with the highest levels of social disadvantage, is one of significant inequality. Municipal governments are generally reluctant to analyze and reconsider, from a local perspective, the disadvantaged conditions in which women live. Local customs and a strong presence of machismo heavily influence local policies in this area.

SDG 6, titled "Ensure availability and sustainable management of water and sanitation for all," includes 8 targets and 12 indicators. Of the indicators presented in this section, only two are directly related to people's access to water; the rest are water management indicators, which are very important for regulatory agencies and service operators. In this regard, it is essential to objectively identify the percentage of the municipal population without access to piped drinking water, the existence or absence of municipal recycling systems, the number of wells exploited in each municipality to supply water to the population, the degree of municipal water stress, and the proportion of contaminated water in natural flows (rivers, lagoons, lakes, streams) by municipality. Water management is becoming one of the main areas of concern at the municipal level. National legislation insists on maintaining a federal monopoly over water administration, centralizing decisions without addressing the specific realities of municipalities and localities. There are black markets for water, speculation and trading occur, often with the complicity of local authorities, resulting in high costs for families and their well-being.

SDG 7, titled "Ensure access to affordable, reliable, sustainable, and modern energy for all," includes five targets and five indicators, none of which are related to people's access to energy sources in their homes. Once again, these are management indicators for the state-owned company responsible for the sector. It is crucial to have accurate data on the coverage of electricity and gas services in households in each municipality, the percentage of households using alternative and lower-quality energy sources, and the percentage of the population opting for other supply alternatives, especially non-polluting sources.

SDG 8, titled "Promote sustained, inclusive, and sustainable economic growth, full and productive employment, and decent work for all," has 14 targets and 30 indicators. In addition to targets and indicators related to per capita GDP growth, average income of employed persons, unemployment rate, youth not in employment or education, informal employment in the agricultural sector, and child labor, this SDG also includes variables related to the strength of local economic activity, such as the number of people with a bank account by gender and age, and the number of ATMs per 100,000 inhabitants. It also reports on the number of employed persons with disabilities, the number of work-related injuries, among others. It should be noted that these indicators are not presented at the municipal level. In this regard, in addition to the variables provided by economic censuses at the municipal level—such as total paid staff, total remuneration, gross municipal value added, and number of economic units—it is essential to know the number of people with savings accounts and the total amount of private municipal savings, as key variables for assessing endogenous municipal development.

SDG 9 refers to "Build resilient infrastructure, promote inclusive and sustainable industrialization, and foster innovation," with 9 targets and 16 indicators, though the criteria for including some of these indicators are not entirely clear. In addition to the variable measuring the percentage of the population living within two kilometers of an all-weather road, there are indicators on passenger transport volume, freight transport volume, the value of manufacturing within GDP, total jobs generated by manufacturing, the proportion of small businesses that have obtained a loan, the percentage of the population employed in small and medium-sized enterprises, carbon dioxide emissions, federal spending on scientific research and experimental development, the multiplier effect of research and experimental development subsidies provided by companies, and the proportion of the population covered by mobile networks, among others.

The key aspects of municipal infrastructure associated with well-being include water distribution and sewage systems for 100% of the population, wastewater and industrial effluent treatment systems, a central bus station within the municipality, rural roads that facilitate connectivity for communities and individuals, primary and secondary care hospitals for the population, quality public schools, municipal internet connectivity for schools, public agencies, businesses, and the general population, and vocational training centers.

SDG 10, titled "Reduce inequality within and among countries," includes 6 targets and 9 indicators. Among these are the proportion of people living below 50% of the median income, disaggregated by sex, age, and disability status; the percentage of the population reporting having felt discriminated against; the share of GDP labor income comprising wages and social protection transfers; the Gini index with and without transfers; the percentage growth of household current income by decile; the share of total income represented by the lowest income earners within the highest income brackets; and the proportion of people over 18 with diverse sexual orientations who have concealed their orientation for any reason. Municipal inequality within Mexican states is highly pronounced. There is no clear definition of the content of this goal, nor is there justification for the inclusion of the selected indicators. According to the Municipal Institutional Capacity Index, out of 2215 municipalities studied, only 73 have high or very high capacity development, while 1425 have low or very low institutional capacities (Huerta Cuervo y Murillo Tovar 2025). Inequality is associated with institutional capacities and not only with differences in family income

SDG 11, titled "Make cities and human settlements inclusive, safe, resilient, and sustainable," comprises 9 targets and 16 indicators. In addition to some indicators that CONEVAL includes in the social lag index and poverty data, it covers the proportion of the population with easy access to public transportation, disaggregated by sex; average commute time by type of locality and population status; the ratio of land consumption rate to population growth by state; the proportion of cities with direct civil society participation structures in urban management; total per capita public spending on environmental protection; the number of people killed or missing due to disasters; the percentage of urban solid waste collected and properly disposed of; the average level of total suspended particulates in cities; and the per capita area of green spaces by state, among others. Here again, management and outcome indicators are presented without distinction. For this SDG, it would seem necessary to include public safety indicators by type of crime at the municipal level.

SDG 12, "Ensure sustainable consumption and production patterns," reports 5 targets and 8 indicators, including per capita domestic consumption, the percentage of urban solid waste collected and recyclable, the incorporation of accounting tools to monitor the economic and environmental aspects of tourism sustainability, and the percentage of municipalities with proper solid waste disposal. Neither the concept of the goal nor the associated variables are clearly defined.

SDG 13 refers to "Take urgent action to combat climate change and its impacts." It again reports the number of people killed or missing due to disasters per 100,000 inhabitants, greenhouse gas emissions by sector, and the excess of suspended particulates and ozone. Only 3 targets and 4 indicators are defined. If livestock are the main source of greenhouse gases, it is essential to provide information on the number of cattle per municipality, the management of their solid and liquid waste, as well as training programs on nutrition and food consumption.

SDG 14 aims to "Conserve and sustainably use the oceans, seas, and marine resources for sustainable development." The indicators proposed to evaluate this SDG are on mangrove areas, and the proportion of the scientific research and experimental development budget allocated to the Navy. There is no information on beach quality, the types of pollutants that they receive, or the pressure from fishing in Mexico's seas.

SDG 15, titled "Protect, restore, and promote sustainable use of terrestrial ecosystems, sustainably manage forests, combat desertification, halt and reverse land degradation, and halt biodiversity loss," reports 5 targets and 9 indicators. These include forest area as a proportion of total land area, the proportion of important sites for terrestrial and freshwater biodiversity that are protected areas, the proportion of protected areas relative to total territory, wetland areas—especially as bird habitats—important biodiversity sites such as mountains, the green coverage index in mountains, and the proportion of endangered species. There is no data on the square meters of green areas per person in each municipality.

SDG 16 seeks to "Promote peaceful and inclusive societies for sustainable development, provide access to justice for all, and build effective, accountable, and inclusive institutions at all levels." It defines 10 targets and 36 indicators, of which 14 refer to crime incidence, 13 to levels of trust in authorities and satisfaction with public services, and the remaining indicators address the representation of specific groups in public administration and the judiciary, as well as the proportion of children registered before age five. Knowing the number of corruption complaints in municipalities, the education level of municipal employees by municipality, the percentage of complaints addressed and resolved, and compliance with budget transparency obligations are fundamental in this section.

Finally, SDG 17 aims to "Strengthen the means of implementation and revitalize the Global Partnership for Sustainable Development." It includes 9 targets and 12 indicators of very diverse nature, for example: total revenue as a share of GDP, the proportion of the national budget financed by domestic taxes, the volume of remittances, the share of debt service relative to exports, access to fixed internet services per 100,000 inhabitants and according to its speed, the proportion of people using the internet inside and outside the home, the fiscal autonomy of subnational governments, growth of the jobs in the cultural sector, the percentage of global and regional indicators available in public information, the gross census value added of the cultural sector, and the percentage of the budget represented by INEGI expenditures. In this SDG, as in others, the concept that the indicators aim to reflect is not clear.

The Mexican government platform, while providing relevant elements, still has not assimilated the shift from the MDGs to the SDGs, as the information available there offers little insight to citizens about the realities that need to be transformed and the gaps and challenges that must be overcome at the local level to achieve the defined goals and objectives. It is essential to reach a basic agreement on the concepts associated with each SDG and the key indicators needed to evaluate each one, as well as to distinguish between management and outcome indicators and to present them at the municipal level.

Since the adoption of the SDGs in September 2015, studies have been conducted exploring the capacities and challenges of each country, analyzing the necessary institutional strategies, along with mapping analyses and projections of results (House of Commons 2016; Lucas et al. 2016; ICSU and CISS). Some of these studies acknowledge the shortcomings and limitations of the SDGs, but at the same time have explored the potential they offer for specific development strategies, as well as the political, social, and institutional changes needed to move toward more sustainable societies (Gómez 2017).

The world was not on track to achieve the 17 Sustainable Development Goals before COVID-19 struck, and now the challenge has multiplied many times over. The COVID-19 pandemic has slowed, and in most cases reversed, the hard-won progress toward the 17 SDGs. Its most alarming consequence is likely its impact on global health (SDG 3), as it has caused 774,291,287 cases of illness and the deaths of more than 7 million people as of January 2024 (WHO 2024). But the pandemic has had a cascading effect on all the SDGs. The global economic recession led to the loss of 255 million jobs, it is estimated that nearly 100 million people were pushed into extreme poverty (SDG 1), and it severely affected the 1.6 billion people working in the informal economy (SDG 8), without social (SDG 1) or health coverage (SDG 3). It has caused an increase in hunger (SDG 2), child labor (SDG 16), and gender inequalities (SDG 5). And although the reduction in human activity provided a temporary respite for threatened plant and animal species (SDGs 14 and 15) and for climate change (SDG 13), it was too brief to have a positive impact on ecosystems. (Focus 2021; Schady et al. 2023)

It is in this context that the National Polytechnic Institute, through the Economic Development Network, invites the reader to critically review and discuss how Mexico has progressed in meeting the SDGs, as well as the future challenges and perspectives. The eleven chapters of this book refer, from different theoretical and methodological perspectives, to the progress and challenges for the fulfillment of the SDGs.

References

Alleyne G, Beaglehole R, Bonita R (2015) Quantifying targets for the SDG health goal. Lancet 385(9964):208–209. https://doi.org/10.1016/S0140-6736(14)61655-X

Birendra KC, Dhungana A, Dangi TB (2021) Tourism and the sustainable development goals: stakeholders' perspectives from Nepal. Tour Manag Perspect 38(April):100822. https://doi.org/10.1016/j.tmp.2021.100822

Consejo Nacional de Evaluación de la Política de Desarrollo Social (CONEVAL). (2019). Metodología para la medición multidimensional de la pobreza en México (3ª ed.). https://www.coneval.org.mx/InformesPublicaciones/InformesPublicaciones/Documents/Metodologia-medicion-multidimensional-3er-edicion.pdf

Consejo Nacional de Evaluación de la Política de Desarrollo Social (CONEVAL). (2020). Fichas iniciales de monitoreo y evaluación 2019–2020 de los programas y las acciones federales de desarrollo social. https://www.coneval.org.mx/InformesPublicaciones/InformesPublicaciones/Paginas/Publicaciones-sobre-Evaluacion-y-monitoreo.aspx

Consejo Nacional de Evaluación de la Política de Desarrollo Social (CONEVAL). (2023). Pobreza en México: Resultados de pobreza en México 2022. https://www.coneval.org.mx/Medicion/MP/Documents/MMP_2022/Pobreza_multidimensional_2022.pdf

Fukuda-Parr S (2016) From the millennium development goals to the sustainable development goals: shifts in purpose, concept, and politics of global goal setting for development. Gend Dev 24(1):43–52. https://doi.org/10.1080/13552074.2016.1145895

Gobierno de México. (2023). Objetivos de desarrollo sostenible: Indicadores. INEGI; Secretaría de Economía. https://agenda2030.mx/#/home

Gómez Gil, C. (2017). Objetivos de desarrollo sostenible (ODS): Una revisión crítica. Papeles de Relaciones Ecosociales y Cambio Global, 18(140), 107–118

Government of the Republic (2018a) Voluntary national review for the high-level political forum on sustainable development. In: Foundations and principles in Mexico for a long-term vision of sustainable development. Progress in Achieving the 2030 Agenda, p 162

Hansson S, Arfvidsson H, Simon D (2019) Governance for sustainable urban development: the double function of SDG indicators. Area Dev Policy 4(3):217–235. https://doi.org/10.1080/23792949.2019.1585192

Huerta Cuervo, Rocío y Murillo Tovar, Luis Pablo (2025). Capacidades institucionales municipales, un índice para comprender el desarrollo local. Análisis Económico, 40(104), 107-125.https://doi.org/10.24275/uam/azc/dcsh/ae/2025v40n104/Huerta

Hansson, S., Arfvidsson, H., & Simon, D. (2019). Governance for sustainable urban development: The double function of SDG indicators. Area Development and Policy, 4(3), 217–235. https://doi.org/10.1080/23792949.2019.1585192

Instituto Nacional de Estadística y Geografía (INEGI). (2023). Objetivos de desarrollo sostenible: Indicadores. Gobierno de México. https://agenda2030.mx/#/home

Kemp, R., Parto, S., & Gibson, R. (2005). Governance for sustainable development: Moving from theory to practice. International Journal of Sustainable Development, 8(2), 12–30. https://doi.org/10.1504/IJSD.2005.007372

Secretaría de Gobernación. (2019). Decreto por el que se reforman, adicionan y derogan diversas disposiciones de los artículos 3o., 31 y 73 de la Constitución Política de los Estados Unidos Mexicanos, en materia educativa. Diario Oficial de la Federación. https://www.dof.gob.mx/

United Nations (1995) International Conference on Population and Development (1994. Cairo). In: Report of the International Conference on Population and Development: Cairo, 5–13 September 1994. United Nations

United Nations (1996) Human development report 1996. España. Ediciones Mundi-Prensa Libros

United Nations (2000) La Cumbre del Milenio: Objetivos, metas e indicadores. Guía del Secretario General (Anexo A/56/326). Naciones Unidas

United Nations Development Programme (UNDP). (2009). Handbook on planning, monitoring and evaluating for development. https://mandeguidelines.iom.int/sites/g/files/tmzbdl2306/files/2023-03/pme-handbook.pdf

United Nations (2015) Transformar nuestro mundo: La Agenda 2030 para el desarrollo sostenible (Resolución A/RES/70/1).https://docs.un.org/es/A/RES/70/1

United Nations (2024) The sustainable development goals report 2024. United Nations Department of Economic and Social Affairs. https://doi.org/10.18356/9789213589755

Municipal Institutional Capacities and Sustainable Development Goals: An Indissoluble Binomial

2

Rocío Huerta Cuervo, Evelia Rojas Alarcón, and Luis Pablo Murillo Tovar

Abstract

The objective is to demonstrate the dependency relationship between the number of poor people in each municipality and the index that measures municipal institutional capacities (MICI). Using a quantitative strategy, a descriptive analysis of the evolution of municipal poverty in Mexico (SDG 1) from 2010 to 2020 was conducted. A linear regression model was constructed using the ordinary least squares method, using the number of poor people and public service coverage as dependent variables (with normalized data) and the MICI. The results corroborate a significant dependency relationship between these variables. This should translate into an effort by relevant stakeholders to develop strategies and implement public and private programs that develop and consolidate municipal institutional capacities, especially those that can improve the well-being of the population. The advancement of the SDGs requires the strengthening of institutional capacities.

Keywords

Sustainable development goals · Endogenous development · Municipal institutional capacities

2.1 Introduction

The Sustainable Development Goals (SDGs) are the result of global commitments that require governments, private organizations, higher education institutions, citizens, and their organizations to work toward eliminating the causes of poverty and environmental degradation, creating conditions for the well-being and development of people worldwide (United Nations 2015; Alleyne et al. 2014; Graute 2016; Kc et al. 2021).

The transition from the Millennium Development Goals (MDGs) to the SDGs involved a critical reflection on measurement systems for tracking progress toward committed goals and targets. While the MDGs "helped mobilize resources and political will… they reduced complex issues to one-dimensional measurement questions" (Hansson et al. 2019, p. 218). The evaluation systems of various governments emphasized formal processes, neglecting

R. H. Cuervo (✉)
Centro de Investigaciones Económicas, Administrativas y Sociales, Instituto Politécnico Nacional, Alcaldía Miguel Hidalgo, Ciudad de México, Mexico
e-mail: rhuerta@ipn.mx

Present Address:
E. R. Alarcón · L. P. M. Tovar
Unidad Profesional Interdisciplinaria de Ingeniería y Ciencias Sociales y Administrativas del Instituto Politécnico Nacional, Iztacalco, Ciudad de México, Mexico
e-mail: erojasa@ipn.mx

L. P. M. Tovar
e-mail: lpmurillotovar@gmail.com

P. Pérez-Hernandez and R. Huerta Cuervo (eds.), *Social Well-Being and the Challenges in the Sustainable Development Goals*, Sustainable Development Goals Series,
https://doi.org/10.1007/978-3-032-14294-8_1

comprehensive assessments of real improvements in people's well-being from the places where they live.

In 2015, 17 SDGs and 169 targets were defined to monitor and evaluate local development processes worldwide. For this purpose, the Technical Committee was established in Mexico specialized in the SDGs (CTEODS), under the federal government (Government of the Republic 2018). By June 2018, all 32 states of the Republic had established SDG Monitoring and Implementation Bodies (ISIODS), and 300 municipalities had joined these activities. Although the cited report (Government of the Republic 2018) identified challenges to advancing SDG implementation—including "ensuring greater inter-institutional coordination, both vertically and horizontally; institutionalizing mechanisms for effective participation and co-creation of public policies; consolidating the ownership of the Agenda by different sectors of society; expanding communication and outreach strategies for the Agenda; training stakeholders at all levels of government; ensuring monitoring and review mechanisms for actions and public policies that transcend administrative terms; and establishing long-term planning with intermediate objectives" (Government of the Republic 2018, p. 21)—it is clear that all of this requires a political and social environment of commitment among political forces, business leaders, and citizens. Situations such as a weak rule of law, violence, and ongoing conflict do not provide the best scenario for advancing SDG implementation at the municipal and local levels. The structural lag in the institutional capacities of municipalities, particularly in local governance, poses a significant constraint to progress and achievement. Currently, the federal government, with the support of INEGI (2023), has a platform called "Sustainable Development Goals," which reports on indicators associated with the SDGs. This platform includes both management and outcome indicators. At the municipal level, the platform provides information on 11 of the 169 targets committed to for monitoring (SIODS 2023).

The objective of this chapter is to analyze the relationship between SDG progress and municipal institutional capacities through a quantitative exercise, using official municipal-level data sources. The number of people living in poverty per municipality is identified as the dependent variable of the Municipal Institutional Capacity Index (ICIM) (Huerta and Vanegas 2020; Huerta and Murillo 2024). The ICIM is a theoretical and conceptual construct that allows for a comprehensive measurement of the institutional strength of municipalities and is associated with three dimensions: economic, governmental, and social.

The results show a significant negative relationship between the level of municipal institutional capacity development and the number of people living in poverty in each municipality. Furthermore, this institutional weakness has fostered phenomena such as violence, which introduces additional challenges to meeting the SDG targets.

Although the chapter references all 17 SDGs, the quantitative study focuses exclusively on SDG 1. This choice is based on the consideration that SDG 1 is a central goal and is interrelated with other key objectives. Addressing it can generate positive impacts, as poverty reduction has a multiplier effect that improves access to quality education, the creation of decent jobs, and access to health services. This work is of particular interest to governments, especially subnational ones, which face persistent challenges in their territories. The structure of this work, following the introduction, is as follows: Sect. 2.2 presents the theoretical frameworks on endogenous development, the SDGs, and new public governance. Section 2.3 provides an initial overview of the diverse realities of Mexican municipalities and explains the Municipal Institutional Capacity Index. Section 2.4 details the data sources and research methodology used, and Sect. 2.5 presents the results and findings, followed by the conclusions.

2.2 Endogenous Development, SDGs, and New Public Governance

2.2.1 Endogenous Development

The twenty-first century is witnessing a growing awareness of the need to address global challenges from local perspectives. In this context, local development theories have emerged as tools to promote sustainable progress by making significant contributions to the achievement of the United Nations Sustainable Development Goals (SDGs).

Local development involves an approach that contrasts with more centralized and homogeneous visions of development, recognizing that a single strategy does not fit all geographic realities. On the contrary, it acknowledges the diversity of contexts and the importance of community participation in the planning and implementation of projects based on the identification of specific internal resources (Boisier 1999). Local development theories aim to generate solutions tailored to the specific realities of each place. Therefore, integrating various theories of local development can offer a robust, holistic, and adaptive option for addressing the diverse challenges posed by the 2030 Agenda.

Achieving the SDGs requires a development approach based on the efficient identification and use of natural, human, and cultural resources in a given region, whether urban or rural. Maximizing the sustainable use of local resources can drive economic and social development (Pérez 2006). In particular, the SDGs offer local governments an opportunity to shift perspectives by focusing on new notions of development, in which economic indicators, while substantial, cannot be the sole measure of population well-being. Local development and the SDGs share a vision centered on improving people's quality of life. They consider aspects such as education, health, and citizen participation with the aim of achieving inclusive and equitable development that directly enhances the well-being of the local population (Sen 1998).

In this regard, Goals 11 (Sustainable Cities and Communities) and 16 (Peace, Justice, and Strong Institutions), emphasise on two fundamental aspects: first, identifying the growth patterns that have developed in large cities and smaller localities; and second, addressing the most undesirable outcomes of these processes through public policy and citizen participation. Large cities, with their more developed infrastructure for health services, education, and greater employment opportunities, have acted as magnets attracting residents from less developed areas. According to the UN report (2023), it is estimated that by 2050, 70% of the world's population will live in urban areas, and of these, more than three billion will reside in marginalized zones, leading to significant challenges in the provision of public services, substandard housing, overcrowding, domestic and community violence, lack of accessible and sufficient transportation services, scarcity of urban green spaces, increased pollution, and limited opportunities for formal employment and decent wages (p. 2).

2.2.2 Theoretical Perspectives on Local Development and Its Relationship with the SDGs

There are various theories of local development that address different aspects and approaches to promoting sustainable progress at the community level. In this context, endogenous development emerges as an organic approach aligned with the SDGs. Endogenous development focuses on leveraging a community's internal resources to drive its own development, strengthen local capacities, and foster community participation in decision-making (Arocena 1995). This approach enables communities to become active agents in their own development, promoting sustainability, equity, and resilience at both local and global levels.

A close relationship between endogenous development and the SDGs becomes evident when considering the intrinsic capacity of some communities to drive their own sustainable growth. By applying endogenous strategies, communities can improve food security (SDG 2) through sustainable agricultural practices adapted to local resources. At the same time, by seeking to reduce dependence on external resources, SDG 1 (no poverty) and SDG 8 (decent work and economic growth) are addressed through the identification of local resources and the comparative advantages of each municipality, opportunities for cooperation and synergy-building with neighboring municipalities, the creation of businesses and enterprises as sources of wealth and employment, professional training, and the promotion of local economic activities. Furthermore, empowering communities through endogenous development processes directly contributes to SDG 5 (gender equality) by enabling more equitable participation of men and women in local decision-making.

From the perspective of local development, the theory of local innovation systems can complement a new notion of development that enables the design of strategies for achieving the SDGs, stimulating creativity, the application of local knowledge, and the development of adaptive capacities within communities. By fostering interaction and collaboration among local businesses, educational institutions, and governments, innovation in key sectors can be promoted, directly contributing to SDG 9 (industry, innovation, and infrastructure). Likewise, this approach can help strengthen the local economy and align with SDG 8 (decent work and economic growth) by creating favorable conditions for job creation, stimulating savings, competitiveness, and local and regional investment (Romer 1992, 1994).

The theory of social capital also plays a crucial role in building an integrated strategy that links the SDGs to local development. For local development, it is necessary to identify and understand the interactions and connections among individuals or groups, as well as to strengthen trust and cooperation between individuals and communities in order to address local problems and seize opportunities. This can be related to and contribute to several SDGs, including SDG 16 (peace, justice, and strong institutions) by fostering cooperation and social cohesion. Additionally, improving citizen participation contributes to the achievement of SDG 11 (sustainable cities and communities) by involving institutions and citizens in planning and decision-making for local development (Sen 2000).

There are other theoretical perspectives on local development that can be considered key components of a strategy for achieving the SDGs by 2030. Territorial development theory analyzes the specific characteristics of a territory with the aim of designing development processes that enhance the conditions of an area and its resilience to physical features and recurring natural phenomena. This territorial analysis involves selecting and distributing the right economic activities to improve the competitiveness and economic efficiency of a region by leveraging local advantages and resources. Factors such as infrastructure, labor force, and natural, physical, and environmental conditions are therefore considered (Barquero 2007; Rodríguez and Caldera 2013). This territorial perspective includes elements such as "…security, fragility, conflict, violence, and a place's vulnerability to natural disasters. The low availability of open spatial data results in lost opportunities for informed policy-making, scientific advancement, and the improvement of social and economic conditions" (Ivanescu and Mayén 2024, p. 5).

These theories related to local development are not mutually exclusive and often complement each other. The choice of theory or approach will depend on the specific characteristics of the community and the development objectives identified as priorities. Ultimately, each of the SDGs addresses specific challenges, and local development theories provide fertile ground for addressing these goals in a contextualized manner. SDG 4, which seeks to ensure quality education, benefits from strategies that incorporate global scientific advances in understanding the universe, society, and technological development into human capital formation processes, and that address the specific needs of students in a given region by incorporating culturally sensitive approaches and integrating local realities into educational programs (Menoyo 2020).

For example, in marginalized areas of India, local development programs have empowered women through skills training and the creation of cooperatives, directly contributing to SDG 5 by promoting equity in political and economic participation. In Africa, community initiatives focused on analyzing consumption patterns and the local availability of water resources have improved access to safe drinking water and sanitation (SDG 6), demonstrating that local solutions for sustainable water management can be effective in addressing global challenges. In rural communities in Latin America, the implementation of agricultural projects based on local resources has not only improved food security (SDG 2) but also fostered sustainable economic development (SDG 8) (ECLAC 2018).

Local development theories face challenges that may limit their effective implementation, such as the heterogeneity of communities and municipalities, lack of financial resources, resistance to changing entrenched practices, and lack of coordination. Additionally, adapting local strategies to highly populated urban contexts presents further challenges. Local governments are closest to the people and therefore may have the means to deeply understand the specific problems and needs of their communities by implementing local development initiatives that allow for greater adaptability to local needs and enable progress assessment, facilitating alignment with the SDGs.

Collaboration among local governments, non-governmental organizations (NGOs), the private sector, and communities is crucial for the successful achieve the SDGs through synergies among stakeholders. Progress toward these goals requires building local governance systems in which the government sector is strongly supported by private and social actors. Local governments can facilitate the creation of policies that support local development initiatives, build capacity at the municipal level, and foster active collaboration among municipalities (Cabrero 2005). In turn, NGOs and the private sector can provide resources and technical expertise as key elements to ensure effective and sustainable development (López 2013).

2.2.3 The SDGs: Toward a New Public Governance

The Millennium Development Goals provided important lessons that have informed the construction of the SDGs, especially in the field of public governance. It is necessary to move from performance-based MDGs to the SDGs as a management tool that contributes to governance for sustainable development. Setting comprehensive development goals for people, starting from their localities, cannot be done without their involvement. The new public governance means engaging the key actors in a community's life in the processes of diagnosis, planning, formulation, implementation, and evaluation of public policies designed to address the most relevant local issues (Fukuda-Parr 2016).

The success of the SDGs lies in the process of local development, which in turn is strongly linked to its comprehensive nature, within which the exchanges and interactions that enable a global vision of local society are generated.

Local societies are subject to constant changes in their environment, such as national policies, fluctuations in international markets, and technological transformations, all of which decisively influence the development of local areas. Not all local societies respond in the same way to environmental changes; their capacity to respond depends on the characteristics of each region. Therefore, it is important to properly implement innovative indicators aimed at changing behaviors that lead to governance for sustainable development, rather than using them performatively to create an impression of progress.

As Aguilar (2022) notes, the new public governance must be based on three key elements: first, a regulatory system conducive to development that encourages wealth generation, substantive inclusion, poverty reduction, and environmental protection; second, a system of scientific and technological knowledge that contributes to understanding the causes of public problems and how they manifest in each locality; and third, a system of political relations that integrates all relevant actors in a region into decision-making processes. At the center of these elements is

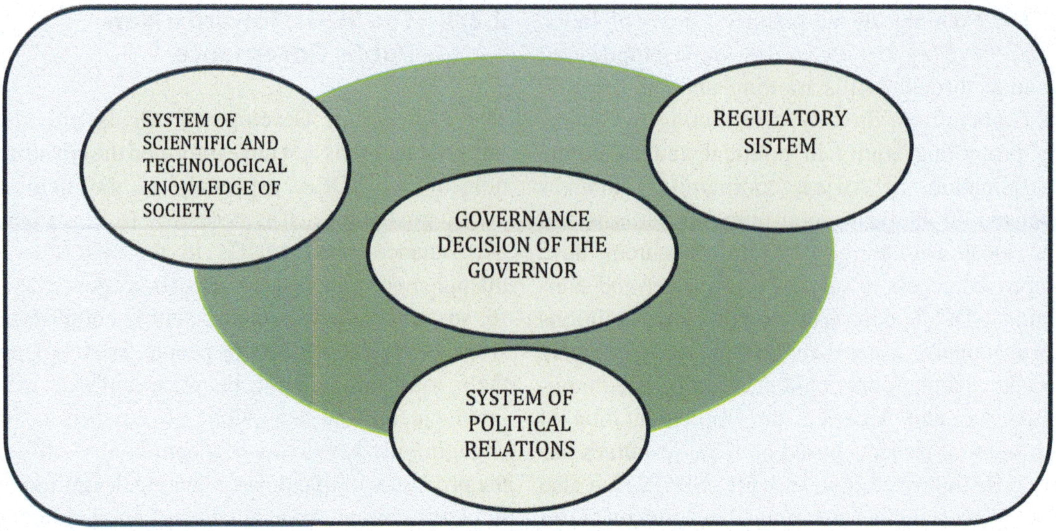

Fig. 2.1 The new public governance. *Source* Own elaboration based on Aguilar (2022)

the exercise of responsible governance by those who govern each municipality in the country (Fig. 2.1).

The transition from the MDGs to the SDGs represents a qualitative shift in understanding the processes required to transform local realities. Thus, the indicator system for providing information to individuals, organizations, and decision-makers must be able to express the specific realities of each jurisdiction. A major challenge for municipal governments and academic institutions in municipalities is building an indicator system that comprehensively reflects the condition and progress of people's well-being. Entities and municipalities have deep differences and areas of opportunity that are difficult to overcome without leveraging scientific knowledge about the causes of these problems.

Data from various indicators relevant to the analysis of a jurisdiction, when placed in the hands of local policymakers, take on new significance as they are enriched by direct and qualitative knowledge of the local reality. This is a primary element that national, state, and municipal coordinators of SDG monitoring processes must address. Indicators for transforming local realities must be grounded at the municipal level. Moreover, not all indicators serve the same function: while management indicators provide valuable information to government areas responsible for certain issues, outcome indicators are vital for citizens and evaluators of SDG progress, as they allow for the identification of the specific status of a variable at a given point in time.

That aspect—the exercise of responsible governance in accordance with regulatory frameworks and knowledge that enables effective intervention in problems—has been a critical factor for municipal governments in Mexico. According to the organizational suitability indicator, obtained in this study based on the Census of Municipal Governments and Territorial Demarcations of Mexico City (INEGI 2021), the national average for this indicator is 0.41 on a scale from 0 to 1, with only 61 municipalities possessing the necessary organizational structures to fulfill the various powers and obligations established by the Political Constitution of the United Mexican States (Fig. 2.2).[1]

[1] ongreso de la Unión (2025). Constitución Política de los Estados Unidos Mexicanos. Consulted in: https://www.diputados.gob.mx/LeyesBiblio/pdf/CPEUM.pdf.

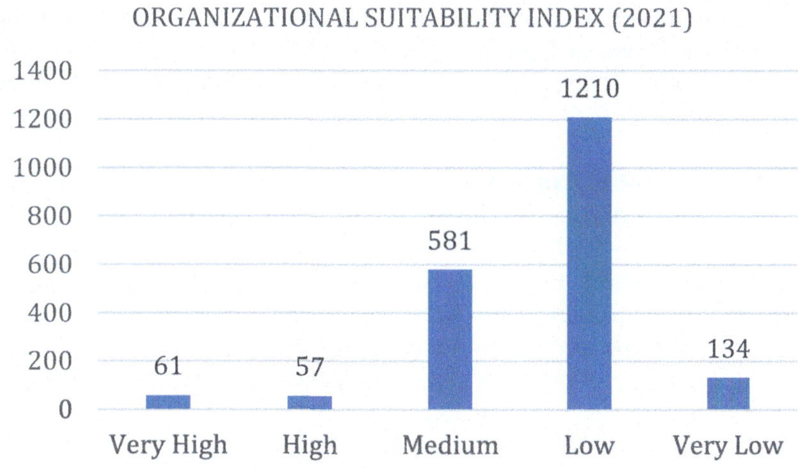

Fig. 2.2 Organizational suitability index 2021. *Source* Prepared by the authors based on INEGI (2021)

2.3 The Municipalities of Mexico: A Diverse Mosaic of Capacities and Resources

2.3.1 The Municipal Institutional Capacity Index (ICIM)

The Municipal Institutional Capacity Index emerged as a result of significant reflection within academic debate, particularly regarding the constraints that public policies have faced in responding to the diverse realities of Mexican municipalities (Oates 2003; Cabrero, 2005; UNDP 2010; Barcelata 2010; Weingast 2009). Although municipal governments were recognized as a third level of government in the 1917 Political Constitution of the United Mexican States, it was not until 1999 that they were granted significant powers, which greatly limited their development and, consequently, their professional capacities (Cabrero, 2005). Each municipality is different and exhibits varying levels of economic, governmental, and social development; therefore, designing national policies based on the premise that all local governments can implement them in a similar manner has been a major constraint to their success. Only recently have municipal governments begun to play a more prominent role, though they remain subordinate to centralized decisions that repeatedly condition their performance. As Meza (2015, p. 133) observed, "local policy agendas are subject to the use of certain instruments by higher levels of government, both federal and state, and therefore do not depend exclusively on local actors and factors." Theoretically, municipal governments are best positioned to respond to and address the specific needs and conditions of the inhabitants of a territory, as they are the level of government closest to the population. However, this general assumption does not fully materialize, since territorial proximity is often not accompanied by the resources and municipal capacities needed to generate relevant diagnoses and effective mechanisms for public and social intervention (Huerta 2015). The causes of public problems are always specific, occurring in a particular space and influenced by a complex combination of factors. Municipal institutional capacities are understood as the set of economic, organizational, and social assets present in a jurisdiction, derived from the operation of a set of historically constructed formal and informal rules (Huerta and Vanegas 2020). "At the core of these capacities is human capital, as the organization and utilization of resources within a jurisdiction depend on it… therefore, the role of municipal governments is crucial in promoting and developing strategies that strengthen these capacities. It is assumed that economic aspects (growth, employment levels, investment, etc.) are not spontaneous products of reality, but rather responses to incentives and deliberate actions by public and private agents" (Huerta 2015, p. 240).

2.3.2 An Initial Approach to Municipal Realities

An initial approach to the diverse realities of Mexican municipalities allows us to broadly identify the significant inequalities present across the national territory, even within states and municipalities themselves. It also highlights the major challenges that the SDGs face in our country and the need to promote differentiated strategies for economic, governmental, and social development.

According to the OECD (2022), Mexico is on track to eradicate extreme poverty (Target 1.1), although 7 million people still live in such conditions. Regarding poverty in general, Mexico records the highest levels of poverty and income inequality (Targets 1.2, 10.1, and 10.2), while redistribution through taxes and transfers is among the lowest in the OECD (Target 10.4). Nevertheless, the trend in this area in Mexico has moved, albeit marginally.

In Mexico, according to the 2019 Economic Census and the 2020 Population Census, and based on data from 2,266 municipalities that provided information on the variables analyzed, the 52 largest municipalities in the country (with more than 500,000 inhabitants) are home to 38% of the total population, produce 51% of the national GDP, account for 38% of economic units, have an average annual formal wage of 127,000 pesos, and provide an average coverage of 67% in the delivery of municipal public services. Approximately 49 million people live in these municipalities, of whom nearly 15 million live in poverty.

In the 180 medium-sized municipalities in Mexico (included in this study), which have between 100,000 and 500,000 inhabitants, 34% of the population is concentrated, 35% of the national GDP is produced, 31% of economic units are located, the average annual salary of formal workers is 104,431 pesos, and municipal public service coverage is 66.2%. Approximately 41% of the population living in these municipalities (17 million people) lives in poverty.

Small urban municipalities, with between 15,000 and 100,000 inhabitants, number approximately 984 in Mexico. Together, they account for 25% of economic units, produce 12% of GDP, and are home to 19% of the total population. The average annual salary of formal workers is 77,883 pesos, and public service coverage is 68.9%. Their population living in poverty is approximately 12 million people.

Finally, the 1,181 rural and semi-rural municipalities contribute 1.4% of GDP, 5.5% of economic units, have an average annual salary of 56,343 pesos, and an average public service coverage of 56%. About 6% of Mexico's total population lives in these municipalities, of which 65% are poor.

The 31 state capital municipalities receive, on average, 80% of federal participations and contributions annually. This helps explain the persistent logic of their growth, but not necessarily the improved provision of goods and services for all their inhabitants (Huerta 2024).

The previous information allows us to broadly identify the main differences among the five groups of municipalities (large urban, medium urban, small urban, semi-rural, and rural). Urban municipalities provide the highest salaries (although these remain quite low on average).

Additionally, a phenomenon that has grown worldwide over the past 20 years is violence and the deaths resulting from it. Intentional homicides cause more deaths than armed conflicts and terrorist attacks combined (UN 2023). Violence by criminal groups and gangs against the population has intensified, especially in Latin America and Africa. Latin America has the highest homicide rate in the world, with 36.6 homicides per 100,000 inhabitants (UN 2023). Drug trafficking continues to draw thousands of young people into drug use, and human trafficking is a growing problem for which there are still insufficient mechanisms for identification, prosecution, and punishment. Actions to redirect the growth processes of urban areas and localities to make them safer and offer a dignified life for all are a central challenge, as is the development of effective policies to combat organized crime groups and gangs that threaten the safety and lives of people.

In 2022, the national crime rate averaged 28,701 offenses per 100,000 inhabitants, with Mexico City having the highest rate at 46,032 offenses per 100,000 inhabitants, followed by the State of Mexico, Querétaro, San Luis Potosí, Puebla, Morelos, Nuevo León, Guerrero, Tlaxcala, and Tabasco (INEGI 2023). These figures illustrate the institutional weaknesses at all three levels of government and the discouraging environment in Mexican municipalities for addressing the challenges of the Sustainable Development Goals.

2.4 Methodology and Data Sources

The ICIM is an index that captures the existing differences among Mexico's various municipalities in economic, governmental, and social dimensions (Huerta and Vanegas 2020; Huerta and Murillo 2024). This makes it possible to confirm the significant disparities among them, their strengths, but also their shortcomings and limitations. The ICIM used 13 variables, which are shown in Table 2.1.

Using the principal component method, a general model was obtained that summarizes the institutional capacities of the municipalities and the indicator value for each municipality. Of the 13 variables used in the calculation of the Municipal Institutional Capacity Index (MICI), the following were significant in the economic component with a coefficient greater than 0.3: municipal GDP, total remunerated personnel, total remuneration, number of economic units, total municipal expenditure, financial autonomy, and population. In the governmental and social component, the weight of public investment and the weight of social expenditure were significant with a coefficient greater than 0.3. Thus, 9 out of the 13 variables considered were significant.

The model representing the Municipal Institutional Capacity Index (MICI) was specified as follows:

$$ICIM = B_0 + B_1C_1 + B_2C_2 + \mathrm{e} \qquad (2.1)$$

2.5 Results

The result of the principal component analysis was:

Table 2.1 Variables comprising the municipal institutional capacity index (ICIM)

Economic component		Governmental component		Social component	
These variables were obtained from the 2019 Economic Census (INEGI)	Municipal Gross Domestic Product (MGDP)	These variables were obtained from the Municipal and Mexico City Public Finance System of INEGI	Total municipal expenditure (TME)	These variables were obtained from the INEGI Population Census (2020), UNDP (2016, 2020), and CONEVAL (2020)	Population (POP)
	Total municipal paid personnel (TMPP)		Municipal financial autonomy (MFA)		Human Development Index (HDI)
	Total municipal remuneration (TMR)		Municipal social expenditure (MSE)		Functional Capacity Index (FCI)
	Average municipal remuneration (AMR)		Average municipal expenditure per capita (AMEPC)		Social Lag Index (SLI)
	Number of economic units (EU)				

Source Own elaboration based on the methodology of Huerta and Vanegas (2020)
Note The abbreviations correspond to those used in the statistical analysis with INFOSTAT software

$$ICIM = B_0 + 0.49C_1 + 0.21C_2 + e \quad (2.2)$$

which means that the economic component carries the greatest weight in the indicator, followed by the governmental and social components. Annex 1 presents the typology of municipalities according to the MICI.

Regarding the SDGs, this study analyzed only SDG 1 (End poverty in all its forms), for two main reasons. In Mexico, the concept of poverty is constructed multidimensionally and includes various indicators that, in the SDGs, appear segmented. The second reason was the space available for the development of this chapter. Evaluating each SDG individually would require more space than is available here.

The information sources for the analysis of SDG 1 are the data provided by CONEVAL for the years 2010 and 2020. The study strategy was quantitative, using linear regression exercises with the Ordinary Least Squares (OLS) method. The number of poor people per municipality (SDG 1) is the dependent variable of the MICI, and the model was specified as follows:

$$ODS1(Pobreza) = B_0 + B_1 ICIM + e \quad (2.3)$$

where SDG represents the data on the poor population per municipality in 2020 and MICI is the value of the index.

Subsequently, a typology of municipalities was constructed based on the level of progress on SDG 1, categorizing municipalities according to convenience criteria.

For the descriptive analysis of poverty, information was obtained from 2,411 municipalities. For the regression exercise between the SDG and MICI variables, data were obtained from 2,182 municipalities that had complete information for both variables.

Poverty is a multidimensional phenomenon (CONEVAL 2019), meaning a lack of basic resources for people to have a decent standard of living and a satisfactory level of well-being. Poverty takes on specific characteristics depending on the territory in which it is analyzed (Katterman, 1998); therefore, municipal-level assessment of poverty is highly relevant and is central to the SDG approach. On average, at the municipal level, the number of poor people in Mexico increased by 6.9% from 2010 to 2020. In 1,108 municipalities, poverty increased, with 26% of people in those municipalities becoming poor. In 1,346 municipalities, the number of people in poverty decreased, on average, by 19%. If in 2010, 52 million Mexicans were poor, by 2020 the figure reached 56,647,995. Notably, in Mexico City, poverty increased in 13 of its 16 boroughs, with half a million more people in poverty from 2010 to 2020 (CONEVAL 2020). Of the 31 state capital municipalities, poverty increased in 14. The belief that it is possible to overcome poverty by migrating to large cities does not seem to hold true in all cases. In capital municipalities such as Campeche, Tuxtla Gutiérrez, Guanajuato, Pachuca, Toluca, Cuernavaca, Tepic, Puebla, Querétaro, Hermosillo, Centro, Victoria, and Xalapa, poverty increased. Table 2.2 summarizes the increase or decrease in the number of poor people between 2010 and 2020 in Mexican municipalities.

Similarly, it is important to note that although the majority of people living in poverty reside in urban municipalities, the percentage prevalence of poverty is higher in rural and semi-rural municipalities. Table 2.3 presents the municipalities with the largest number of people living in poverty in Mexico in 2020. Notably, these include municipalities in the State of Mexico, neighboring boroughs of Mexico City, and municipalities in the metropolitan area of the state of Jalisco.

Poverty is associated with the level of municipal institutional capacity (ICIM). The results of the regression analysis show a significant ($p < 0.001$) and relevant negative relationship between the two variables, with a coefficient of determination ($R2$) of 33%. This indicates that 33% of the variability in poverty levels can be explained by differences in municipal institutional capacities.

As shown in Table 2.4, based on the regression results and the corresponding Fig. 3, for each one-point increase in the ICIM, an average decrease of 14.12% in the number of people living in poverty is expected. This inverse relationship suggests that strengthening institutional capacities at the municipal level has a considerable impact on poverty reduction.

Table 2.2 Municipalities with an increase and decrease in the number of people in poverty between 2010 and 2020

Percentage increase in the number of people living in poverty	More than 100%	Between 50 and 99%	Between 25 and 49%	Between 10 and 24%	Less than 10%	Total municipalities
Number of municipalities where poverty increased	32	106	215	341	369	1063
Percentage decrease in the number of people living in poverty	Less than 100% and more than 50%	Between 25 and 49%	Between 10 and 24%	Between 5 and 9%	Less than 5%	
Number of municipalities where poverty decreased	50	211	199	535	353	1348

*Source*ONEVAL (2020)

Table 2.3 Municipalities with the largest number of people living in poverty in Mexico, 2020

Municipalities with the largest number of people living in poverty in Mexico (2020)			
León	816,934	Naucalpan de Juárez	382,170
Ecatepec de Morelos	786,391	Gustavo A. Madero	381,752
Iztapalapa	772,584	Juárez	379,290
Puebla	680,945	Guadalajara	336,622
Nezahualcóyotl	523,289	Benito Juárez	323,435
Toluca	511,347	Ixtapaluca	319,519
Chimalhuacán	493,687	Valle de Chalco Solidaridad	278,768
Tijuana	453,508	Tlalpan	274,221
Acapulco de Juárez	394,861	Álvaro Obregón	273,656
Zapopan	387,944	Tlajomulco de Zúñiga	271,636

Source Own elaboration based on CONEVAL (2021)

Table 2.4 Regression analysis between the % of poverty and the ICIM

Variable	N		R	Adj. R	RMSE	AIC	BIC	
% Poverty 2020	2182		0.33	0.33	314.53	18,731.44	18,748.5	
Regression coefficients and associated statistics								
S.O.V	SS	df	MS	F	p-value			
Model	337,815.5	1	337,815.5	1080.78	0			
ICIM	337,815.5	1	337,815.5	1080.78	0			
Error	681,395.5	2180	312.57					
Total	1,019,211	2181						
Type III SS variance analysis table								
Coef	Est	S.E	LL (95%)	UL (95%)	T	p-value	CpMallows	VIF
const	60.1	0.38	59.36	60.84	158.79	0		
ICIM	−14.12	0.43	−14.96	−13.27	−32.88	0	1080.779	1

*Source*CONEVAL (2020) and Huerta (2016)

Based on the results, the most relevant component within the ICIM (49%) is the economic component. The value of municipal GDP, total compensation, and the number of economic units have a significant influence on poverty reduction. Direct support to the poor population, which is included in the income variable used by CONEVAL to measure poverty, improves people's conditions in the short term but does not enable them to overcome poverty. In component 2 (governmental—social) of the ICIM, the variables of the share of municipal social spending, which includes municipal public investment and financial autonomy, were more relevant. Both are related to the public services provided by municipalities and are a factor in the population's well-being. It is important to emphasize that effective poverty reduction must impact the structural variables of the economy and the long-term conditions of families.

2.6 Conclusions

Mexico is a country with great cultural, economic, and geographic diversity. As a result, the Mexican municipal context presents a series of unique characteristics and challenges that affect local development and the implementation of public policies. The results show how municipalities vary greatly in economic development, infrastructure, and access to basic services. Given the marked inequality between rural and urban municipalities, rural municipalities tend to face higher levels of poverty and lower institutional capacity. However, poverty persists even in urban municipalities that are state capitals and key entities in the country. This demonstrates that institutional capacity is a common denominator that must be studied in all its dimensions.

The lack of institutional capacity implies constraints in human, financial, and technical resources, which affects the ability of municipalities to efficiently design, implement, and manage action plans and public policies. Aspects such as training and professionalization of municipal staff are a constant need to improve local governance. Likewise, the financial autonomy of municipalities is limited due to economic informality and the lack of efficient collection mechanisms, as well as the fact that most municipal revenues come from federal and state transfers, which can limit the ability of municipalities to design and execute independent local projects, especially those related to the 2030 Agenda and its objectives.

A defining feature of Mexican municipalities is their pronounced heterogeneity. Addressing and confronting the problem of poverty requires not homogeneous national policies, but rather policies that identify the differentiated and specific realities of municipalities in their respective territories. Poverty is significantly dependent on the level of development of municipal institutional capacities. This result leads us to understand that overcoming poverty can only be achieved through public policies that impact people's living conditions at the local level. Developing capacity cannot be the product of a single government or of short-term efforts. It requires medium- and long-term strategies that are sufficiently agreed upon by relevant political, economic, and social actors within the municipalities. Municipal agendas cannot be subordinated to other levels of government in the central interest of certain political actors. Progress on the SDGs requires that municipal government agendas arise from diagnoses based on these realities. This conclusion makes it essential to subordinate the interests of groups and actors to the specific problems and needs of the people living in a given jurisdiction. Prioritizing the development of institutional capacity requires the rule of law, professionalization of municipal governments, and quality education as permanent and necessary inputs.

Acknowledgements This research was funded by the Instituto Politécnico Nacional through the Secretaría de Investigación y Posgrado, project SIP 20240870.

Annex 1

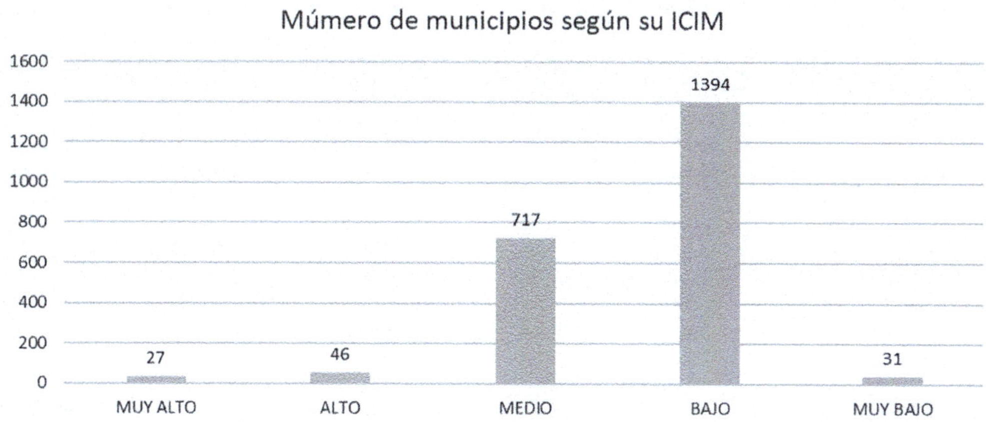

Typology of Mexican municipalities according to the ICIM. *Source* Own elaboration based on ICIM results

References

Aguilar Villanueva LF (2022) About government. theory proposal 1st edn. Tirant lo Blanch. Political Science, 198 pp.

Alleyne G et al (2014) Quantifying targets for the SDG health goal. Lancet 385(9964):208–209. https://doi.org/10.1016/S0140-6736(14)61655-X

Arocena J (1995) Local development: a contemporary challenge, vol 158. Nueva Sociedad, Caracas

Barquero AV (2007) Endogenous development. Theories and policies of territorial development. Investigaciones Regionales = J Reg Res (11):183–210

Barcelata Chávez H (2010) Las finanzas públicas de los estados de México. Análisis Económico, XXV(58), 163–189. https://www.redalyc.org/articulo.oa?id=41313083009

Boisier S (1999) (Local) development: what are we talking about? Santiago de Chile, 28

Cabrero Mendoza E, López L, Segura F, Silva JA (2005) Municipal action and local development: what are the keys to success?

CONEVAL (2021) Plataforma para el análisis terrictorial de la pobreza. Consulted in https://www.diputados.gob.mx/LeyesBiblio/pdf/CPEUM.pdf

ECLAC (2018) Economic commission for Latin America and the Caribbean. 2030 Agenda and the sustainable development. An opportunity for Latin America and the Caribbean, Goals. ISBN: 978-92-1-058643-6 (PDF version)

Fukuda-Parr S (2016) From the millennium development goals to the sustainable development goals: shifts in purpose, concept, and politics of global goal setting for development. Gend Dev 24(1):43–52. https://doi.org/10.1080/13552074.2016.1145895

Government of the Republic (2018) Voluntary national review for the high-level political forum on sustainable development. Retrieved from https://agenda2030lac.org/es/informes-nacionales-voluntarios-inv

Graute U (2016) Local authorities acting globally for sustainable development. Reg Stud 50(11):1931–1942. https://doi.org/10.1080/00343404.2016.1161740

Hansson S, Arfvidsson H, Simon D (2019) Governance for sustainable urban development: the dual function of SDG indicators. Area Dev Policy 4(3):217–235. https://doi.org/10.1080/23792949.2019.1585192

https://www.inegi.org.mx/programas/ccpv/2020/#Microdatos

Huerta Cuervo R (2015) Gobiernos municipales, capacidades profesionales y oferta educativa. En Los desafíos del desarrollo económico local (pp. 237–257). Editorial Porrúa. https://www.researchgate.net/publication/317290443_Gobiernos_municipales_capacidades_profesionales_y_oferta_educativa

Huerta Cuervo R (2016) Educación superior, universidades y gobierno universitario. En R. Huerta Cuervo (Coord.), Construyendo Gobernanza Académica en el IPN (pp. 13–36). Editorial Libermex. https://libermex.com.mx/libreacceso/gobernanza.pdf

Huerta Cuervo R (2024) Municipal public debt in Mexico from 2009–2021: the effects of legal reforms and the COVID-19 pandemic. Revista Finanzas y Política Económica 16(1):95–120

Huerta Cuervo R, Vanegas López M (2020). Methodology for constructing the Municipal Institutional Capacity Index (ICIM). Sobre México. Temas de Economía. Nueva Época 1(2):101–133.

Huerta and Murillo (2024) The municipal institutional capacity index (ICIM) as a tool to explain municipal development processes in the State of Mexico. In: Mexiquense EL (ed) Local dynamics for development, pp 28–54

INEGI (2021) State and municipal public finance statistics 2020. https://www.inegi.org.mx/sistemas/olap/proyectos/bd/continuas/finanzaspublicas/fpmun.asp?s=est&c=11289&proy=efipem_fmun

INEGI (2021) 2020 population census

INEGI (2023) Crime incidence. Government, security, and justice. https://www.inegi.org.mx/temas/incidencia

Ivanescu C, Mayen Huerta C (2024) Can geospatial data help bridge the gender employment gap? yes, it can! [Blog post]. World Bank Blogs. Retrieved from https://blogs.worldbank.org/opendata/can-geospatial-data-help-bridge-gender-employment-gap-yes-it-can

Kattermann D (1998) Institutional analysis for capacity building. In: Borner S, Paldam M (eds) The political dimension of economic growth. International Economic Association Series. Palgrave Macmillan, London. https://doi.org/10.1007/978-1-349-26284-7_16

KC B, Dhungana A, Dangi TB (2021) Tourism and the sustainable development goals: stakeholders' perspectives from Nepal. Tourism Manage Perspect 38:100822. https://doi.org/10.1016/j.tmp.2021.100822

López Oropeza M (2013) Alternative models of local development from the periphery, Eutopía. J Territ Econ Dev 1:115–128. https://doi.org/10.17141/eutopia.1.2010.937

Menoyo MÁM (2020) The road to the SDGs. Comillas J Int Relat 19:1–11. https://doi.org/10.14422/cir.i19.y2020.001

Meza C, Oliver D (2015). Gobiernos locales y agendas de gobierno: ¿a qué responden?. Cuadernos de Trabajo del CIDE. https://repositoriodigital.cide.edu/handle/11651/273

Oates WE (2003) Chapter: an essay on fiscal federalism. In: Baimbridge M, Whyman P (eds) Fiscal Federalism and European economic integration, vol 37, no 3. Routledge, pp 13–47. https://doi.org/10.4324/9780203987254

OECD (2022) The short and winding road to 2020. Measuring distance to the SDG targets, Mexico. OECD Publishing. Available at: https://doi.org/10.1787/af4b630d-en

Pérez MM (2006) Sustainable local development. Economía y desarrollo 140(2)

Rodríguez-González JA, Caldera-Ortega AR (2013) Economic growth and local development in the Centro-Bajío Region of Mexico. Quivera J Territorial Stud

15(1):37–59. Retrieved from https://quivera.uaemex.mx/article/view/9984
Romer PM (1992) Two strategies for economic development: using ideas and producing ideas. World Bank Econ Rev 6(suppl):63–91. https://doi.org/10.1093/wber/6.suppl_1.63
Romer PM (1994) The origins of endogenous growth. J Econ Perspect 8(1):3–22
Sen A (1998) Well-being, justice and the market. Ediciones Paidós Ibérica, Barcelona
Sen A (2000) Development and freedom. Ediciones Planeta, Barcelona
SIODS (2023) Sustainable development goals information system. Mexico. Retrieved from https://agenda2030.mx/#/home
UNDP (2010) Capacity development. Measuring capacity. Capacity Development Group, Bureau for Development Policy, New York
UNDP (2016) Municipal functional capacity index. Retrieved from https://pad.undp.org.mx/#/home
UNDP (2020) Municipal human development index 2020. Retrieved from https://www.idhmunicipalmexico.org/
United Nations (2015), Millennium development goals report 2015, United Nations, New York
Weingast BR (2009) Second generation fiscal federalism: the implications of fiscal incentives. J Urban Econ 65(3):279–293. https://doi.org/10.1016/J.JUE.2008.12.005

3

SDGs, Planetary Citizenship, and Humanistic Education: Gaps and Shortcomings

Napoleón Rosario Conde Gaxiola, Luis Arturo Rivas Tovar, and Angel Eustorgio Rivera González

Abstract

The objective is to analyze the impact of the Sustainable Development Goals (SDGs) and their relationship with Planetary Citizenship and Humanistic Education, which are necessary conditions for their advancement. The research method is hermeneutic, documentary, and heuristic within the three aforementioned categories of analysis. The result is an assessment of the gaps perceived in the context of interpreting and understanding the phenomenon of sustainability, as presented by the 2030 Agenda. As a conclusion to this reflection, the urgent need for a reform of planetary education becomes evident, one that strengthens humanistic education and builds a critical citizenry capable of preventing the destruction of nature, totalitarianism, epistemological and argumentative ignorance, institutional bureaucratization, and colonialism, moving us toward a communal consciousness that distances us from "single-minded thinking" and steers us away from mere hospitable complacency.

Keywords

SDGs · Sustainability · Planetary citizenship · Humanistic education · Gaps and shortcomings of the SDGs

3.1 Introduction

The concern for creating a universal consciousness began with the Universal Declaration of Human Rights in 1948 and the International Covenant on Economic, Social and Cultural Rights in 1966, which would lead to the definition of the Millennium Development Goals (MDGs) in 2000. These goals, enthusiastically approved at the 55th UN Assembly, despite limited achievement, represented a major first step toward building a global agenda that enabled the reduction of poverty and hunger, revived the longstanding aspiration for universal basic education, and promoted the goal of developed countries contributing 0.7% of their GDP to developing countries (United Nations 2023).

Unlike the Millennium Goals, which set out eight major objectives and 18 targets that were difficult to measure due to a lack of data from

N. R. C. Gaxiola (✉)
Escuela Superior de Turismo del Instituto Politécnico Nacional, Ciudad de México, CDMX, Mexico
e-mail: napoleon_conde@yahoo.com.mx

L. A. R. Tovar
Escuela Superior de Comercio y Administración, Unidad Santo Tomás del Instituto Politécnico Nacional, Ciudad de México, CDMX, Mexico
e-mail: larivas33@hotmail.com

A. E. R. González
Unidad Profesional Interdisciplinaria de Ingeniería y Ciencias Sociales y Administrativas del Instituto Politécnico Nacional, Ciudad de México, CDMX, Mexico
e-mail: aerivera@ipn.mx

P. Pérez-Hernandez and R. Huerta Cuervo (eds.), *Social Well-Being and the Challenges in the Sustainable Development Goals*, Sustainable Development Goals Series,
https://doi.org/10.1007/978-3-032-14294-8_2

poor developing countries with significant institutional weaknesses, the new design was innovative. Although the number of goals increased to 17 and targets to 169, these were defined by 230 indicators that could be verified (Weitz et al. 2015).

Global institutions and organizations considered that a significant portion of both the goals and targets were merely rhetorical, and that the targets were more a list of good intentions than truly achievable objectives. Furthermore, the fact that among the signatories were autocratic or outright dictatorial leaders, such as North Korea's Kim Jong-Un and Vladimir Putin, raised well-founded doubts about their fulfillment (Gil 2018; Sarvajayakesavalu 2015; Kroll 2015).

As is well known, the Sustainable Development Goals became the 2030 Agenda, which stands as the most important theoretical and practical framework for creating a paradigm of protection for human rights, nature, the environment, economic and social well-being, poverty reduction, cultural heritage, health, and the prospects for a more hopeful future for the world's population.

The essence of the SDGs lies in the common good, and therefore moves away from instrumentalization and mechanization, so that it becomes not an unattainable utopia, but rather the responsible care of humanity by human beings themselves. On this path, education is necessary to achieve its goals, requiring the construction of a cordial citizenry capable of critically subsuming the proposed dispositions. This chapter aims to interpret the answers to the following questions: What educational paradigm does urban and rural subjects require to realize and successfully guide the aforementioned agenda? What kind of citizenship entails having a border consciousness capable of linking ethics, aesthetics, axiology, and environmental responsibility? What possibilities exist for overcoming the univocal stances typical of the instrumental reason of economic institutions and social formations fetishized by accumulation and consumption? To what extent is it possible to build a critical citizenry in times of war, technocratic hegemony, the predominance of international crime, the growth of military power, hybrid processes of acculturation, and a widespread absence of democratic projects?

Undoubtedly, humanistic education enables the weaving, in a filial and affective manner, of communities that are hermeneutically bearers of an earth consciousness of a planetary, ecumenical, universal, ecological, and political nature. These points—and taking into account SDGs 4 and 16—will be addressed in the following lines, without entirely ignoring the remaining objectives. The aim is not to meticulously address the ten targets and eleven indicators of Goal 4, Quality Education, nor the twelve targets and twenty-three indicators of Goal 16, Peace, Justice, and Strong Institutions, but rather to establish a general overview, taking as a starting point the interaction of education, citizenship, and sustainability in a dialectical and analogical manner.

3.2 Method

This is a hermeneutic investigation of a documentary, bibliographic, and textual nature, complemented by an inductive and deductive methodology that enables a functional articulation between the universal and the particular, form and content, and the axes of problematization and hypothesizing. It also allows us to foster interpretation and understanding over mere description. The treatment of the analytical categories is comprehensive in nature due to their transformational nature. For this reason, we strive to avoid commensurability or univocal measurement of the processes addressed. These are magnitudes capable of possessing a determined value within a dialectical and hermeneutic framework that allows us to move from the abstract to the concrete, and from meaning to reference, enabling a conceptual and thematic shift of a rational and interpretive nature. Dialectics is incorporated as a methodical procedure since it provides the connection between the horizon of contradiction and the framework of proportionality. This formulation historically emerges from the classical world, modernity, and the contemporary era and constitutes an essential artifact in the fabric of the social sciences and humanities. Dialectics, together with the hermeneutic approach, can provide it with a comprehensive and interpretive perspective to establish a structural balance between conceptual reasoning, characteristic of the explanatory model, and the prudential understanding typical of criticism and self-criticism.

3.3 Results

The current situation we are experiencing allows us to visualize global problems that must be addressed immediately to minimize the risks of a global catastrophe. Citizenship and education are essential foundations for recognizing and solving these problems. To address directly the topic of the role of planetary citizenship and education within the SDGs, it is necessary to mention that sustainability is one of the most widely discussed points of reflection in the current context of humanity, given the enormous economic, political, social, pandemic, and environmental crises we have endured over the past three decades.

The sustainable development approach was so significant that it served as the foundation for the MDGs, established through the Millennium Declaration, whose general intention was to improve the living conditions of the world population by 2015. In the same way, they are the immediate precursors of the SDGs and set out the challenges to be addressed in the new millennium.

Figure 3.1 illustrates the difference between the limited Millennium Goals, which, with very narrow targets, sought only to combat extreme poverty and hunger, achieve only primary education, reduce gender inequality, and address very basic health aspects such as improving maternal health, reducing child mortality, and combating malaria and AIDS, and, in a very general way, ensuring environmental sustainability. The mere use of vague and ambiguous verbs, which are more typical of policies than of objectives—such as Promote, Foster, Combat, Reduce, Improve—implicitly favored their noncompliance and provided a ready-made excuse (Rivas-Tovar 2021).

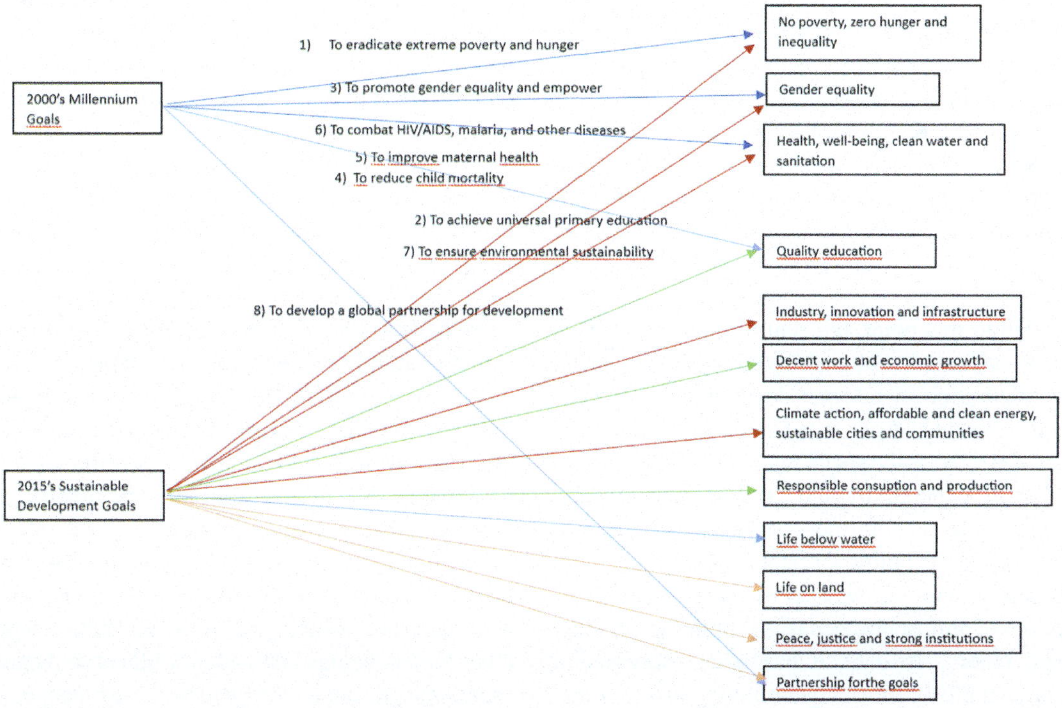

Fig. 3.1 Comparison of the scope of the Millennium Development Goals and the Sustainable Development Goals. *Source* Own elaboration based on United Nations (UN) (2023), Rivas-Tovar (2023)

It is important to note that there has historically been a process of denaturalization of sustainability, citizenship, and education. In this regard, it is feasible for the government and the state to participate and assume responsibility in situations of vulnerability, as sufficient opportunities have not been provided for access to free education, a living wage, housing, healthcare, respect for human rights, nonviolence, and other proposals syntactically outlined by the SDGs. Equal opportunity in priority areas such as health, education, and housing is a pending task for the government systems of many societies.

It would be absurd to discard such a term, but it is necessary to deconstruct it in order to redefine its meaning. The methodology for this purpose is qualitative, hermeneutic, and dialectical, in dialogue with other sciences; therefore, for now, no instrument, variable, or measurement is required—only a set of interpretations morally oriented toward factual transformation. This approach will allow us to visualize not only the consensus and proportions typical of analogy, but also the contradictions, changes, and dynamics inherent in the genesis, development, and evolution of dialectics from a hermeneutic perspective aimed at constructing a syntagmatic and paradigmatic language that allows us to characterize the proposed categories of analysis. In this way, dialectics acquires a dialogical nuance, derived from its connection with hermeneutics, as it transcends supersensible appearance, syllogism, trivium, hasty synthesis, and abstract generalization.

Up to this point, the central question is: How can sustainability be analogized? We approach the configuration of relevant development based on the following ideas: It will be addressed epistemologically by the social sciences and the humanities in general, by the bodies of knowledge that comprise the former—sociology, economics, anthropology, political science, law, ecology, and related fields—as well as the humanistic disciplines of philosophy, pedagogy, philology, history, and others. To date, it has been addressed in some contexts from a quantitative, instrumental, and institutionalist perspective, leading us to view the SDGs from an exclusively descriptive and explanatory standpoint, rather than a phenomenological, hermeneutic, and interpretive one.

The challenges of the twenty-first century—from climate change to ungovernability—require a virtuous articulation between explanation and understanding, natural and cultural knowledge, physicalism and culturalism, mathematical reasoning and anthropological insight, ontology and the establishment of laws, principlism and moral criteria, reason and intuition, the social and the economic, epistemic pluralism and political consciousness. Only in this way can we approach a frontier of sustainability, citizenship, and humanistic education. Human beings must be both concerned and prepared to respond appropriately to the negative impacts of our actions, which have gradually deteriorated and altered the planet's natural cycles and boundaries (Rockström et al. 2009). From our perspective, individualism has been unable to address or resolve these issues. We must therefore also consider elements such as collaboration, trust, and interdependence, since human beings, in addition to being "*homo economicus*" condemned to degrade the planet, are also "*homo reciprocans*" (Fehr and Gächter 1998; Dohmen et al. 2009), capable of feeling and practicing justice, reciprocity, and cooperation in order to find solutions to their problems.

The development of the SDGs is a significant contribution toward reducing or mitigating the most serious current problems within the existing framework of production relations; that is, it is a proposal for social, political, economic, and ecological responsibility. Historically, they have been the backbone of the humanist project to achieve a more livable social formation.

To move closer to achieving the goals of the SDGs, a planetary citizenship is essential—one that can transcend the outdated concept of the nation-state. Such citizenship must be capable of proposing distributive mechanisms for use values and natural products that foster solidarity and reciprocity, which do not yet exist at a universal level. This requires rethinking our relational nature with other human beings and with our planet; it means exercising, through our patterns of exchange, a better form of interaction that allows us to evolve without destruction. It also, requires the eradication of corruption and war, the care of nature and culture, and a commitment that is currently absent from the world's governments. How can poverty be eliminated in the face of the immense concentration of global wealth, where 10% of the population controls 90% of it, while the rest live in precariousness, vulnerability, fragility, and abandonment?

All of this leads us to rescue the SDGs from a univocal model of anthropocentric approach, where the only relevant factor is production—that is, the economy—based on accumulation, profit rate, and profitability. The univocal perspective has been hegemonic in the SDGs and in state-oriented bureaucratic institutions. Some authors refer to this as *very weak sustainability* because it considers only human survival, disregarding other natural and animal elements that surround us. It is a *superhuman* and self-referential stance, lacking ethical criteria and ignoring animal rights, with the *primary analogue* being the general interests of humankind (Pearce and Atkinson 1993, p. 18).

Also part of this trend, with some of its tenets, is Julian Simon through his work *The Ultimate Resource* (1983), in which he emerges as a defender of technology as an alternative to address any environmental conflict.

In the case of environmentalist Kerry Turner (1993), he introduces the concepts of weak and strong sustainability, both fundamental for understanding how humans relate to natural resources. His theoretical position on the matter has shifted as his thinking has evolved. In this context, it is important to define what is meant by strong sustainability, which advocates for preserving natural resources above all else, since failing to do so could compromise the Earth's ability to sustain life, including human life. This is a typical equivocal or relativist position, supportive of an ecologist model. Some of its representatives include: Herman Daly, from the Environment Department of the World Bank in Washington, DC, with his work *Beyond Growth: The Economics of Sustainable Development* (1996, pp. 31–*ff*.); and Tim Jackson, a distinguished professor at the University of Surrey, United Kingdom, with his book *The Earthscan Reader in Sustainable Consumption* (2006), who proposes sustainable consumption as a central concept for moving toward a more balanced relationship between humans and nature.

A position that is neither absolutist nor relativist is that of Mexican environmental sociologist Enrique Leff (1988), who situates sustainability from a critical and proportional perspective. That is, he takes into account the social, cultural, economic, and ecological dimensions in a highly rich and integrated interaction, which we find extremely interesting, as he draws on the German philosopher Martin Heidegger to phenomenologically propose an ontology of life that critically questions technology itself (Leff 2019, p. 223). The Indian economist Amartya Sen is also important. Together with Martha Nussbaum, he has addressed the issue of quality of life, discussing justice, social economy, and the protection of nature from a perspective of equality, equity, and ethical principles. Both scholars, economist and philosopher, agree that sustainability should be linked to equality—gender, economic, social, and ecological. That is, justice in terms of gender, capabilities, traditions, and the political and economic valuation of well-being (Nussbaum and Sen 1996). On the other hand, in his book *The Idea of Justice*, Amartya Sen (2010) proposes a theory of justice with a planetary scope, that is, in line with modern society.

The Spanish hermeneutic philosophers Adela Cortina and Jesús Conill Sancho, along with Uruguayan Gustavo Pereira, in their book *Poverty and Freedom* propose eradicating poverty from the perspective of Amartya Sen's capabilities approach (Conill Sancho 2009; Cortina 2009). They offer an analogical critique of the lack of equality and the limits of liberalism based on the capabilities approach, as articulated by the distinguished Indian Nobel Laureate in Economics, 1998. From another angle, the French sociologist Edgar Morin advocates for the construction of planetary citizenship and a new kind of environmental education. In his book *Flat Earth* he proposes an Earth-based citizenship grounded in our responsibility to the planet. He writes:

> Until the 1950s and 1960s, we lived on an unknown Earth, we lived on an abstract Earth, we lived on an Earth-object. The end of the century revealed to us

> the Earth-system, Gaia, the biosphere, the Earth as a cosmic parcel, the Earth as Homeland. Each of us has our own gnoseology and our terrestrial identity card. We come from the Earth, we belong to the Earth, we are on the Earth. We belong to the Earth, which belongs to us. (Morin 2006, p. 209)

Sustainability in general has been viewed as *techné* or technique, technology, or instruments, based exclusively on *episteme* or knowledge, without connection to *phronesis* or prudence. In proposing *scientia* without virtues and linking it to the State, power, governance, and global domination, while excluding ethical approaches, has become an institutional program that seeks to praise the society of commodities. It has also ignored the contradictions between capital and labor, which make the eradication of poverty impossible, as well as the antagonisms between imperial powers and dependent nations resulting from the fierce competition among imperial blocs such as the United States, European Union, Japan, China, and Russia.

The issue of sustainability is most directly observed in the field of education, addressed in Goal 4. The 2023 report indicates that insufficient progress has been made in terms of graduation rates, free access, gender parity, reading comprehension and math skills, digital literacy, infrastructure, and teacher training, all of which have been further impacted by the pandemic:

> Progress toward quality education was already slower than necessary before the pandemic, but COVID-19 has had devastating effects, causing learning delays in four out of five of the 104 countries studied.
>
> […] Since 2019, UNESCO member states have participated in a national benchmarking process for Goal 4 to set national education targets. Despite the aspiration for universal completion of secondary education, based on their national targets, only one in six countries aims to achieve this goal by 2030. Even if these targets are met, it is estimated that 84 million children and young people will not be in school in 2030. (United Nations 2023, p. 20)

3.3.1 Planetary Citizenship

By this term, we understand the possibility of building a cognitive, democratic, and communal subject endowed with social responsibility in ecumenical terms to reflect, think, feel, act, interpret, and transform the world around us in all its diversity, through economic, political, social, cultural, and ecological recognition, with the purpose of participating through conscious action in their situation and position in relation to reality. It is about transforming the abstract individual into a person capable of leaving behind their consumerist and passive condition, adopting a radical stance against the catastrophe of total democracy, the crisis of governance models, and the lack of solidarity in the face of migration, violence, poverty, rising crime, and other scourges we have experienced.

Planetary citizenship considers human individuality within a constant collectivity; that is, a citizen exists because he or she actively participates, belongs to, and modifies the society in which he or she lives daily. A planetary citizen understands and attempts to resolve social problems, the problems of his or her community (of other humans), but at the same time recognizes the planet as his or her home and tries to avoid, in the best of cases, and remedy, in the extreme case, the problems we have caused to Earth. In this sense, planetary citizenship refers to the relationship that individuals establish with their community, identifying with it and feeling an active part of it (Novo 2006, p. 373).

Although until September 2013 it was still possible to dream of a proposed citizenship that prioritized sustainability, the Russian invasion of Ukraine marked the first war between two nations in decades. The subsequent conflict between Israel and Hamas in the Gaza Strip, since October 7, 2023, which has resulted in more than 23,000 Palestinian deaths, has detracted from the lofty goal of seeking peace (El País 2023); the Russian invasion of Ukraine has resulted in more than 10,000 civilian deaths (EL PAÍS 2023) since it began in February 2022.

In Mexico, the figures resulting from the deadly violence of major drug cartels have led to a death toll far exceeding 134,000 homicides so far in this presidential term. And unfortunately, this is not the only case:

> According to the Geneva Academy of International Humanitarian Law and Human Rights, there are

> 45 active armed conflicts in the Middle East and North Africa, 35 in the rest of the African continent, 21 active conflicts in Asia, 7 in Europe, and Latin America is the scene of six armed conflicts, although these also include criminal violence, as in the case of Mexico.
>
> [...] Also in the north of the continent, Morocco maintains a military occupation in Western Sahara, a territory not recognized as part of the country. Other countries in conflict on the continent include Somalia, Sudan, South Sudan, Cameroon, the Central African Republic, Ethiopia, Mali, Mozambique, Nigeria, and Senegal. (Hernández-Orozco 2023).

In summary, the SDG objectives have not been fully achieved, especially the two highlighted: (4) Quality Education and (16) Peace, Justice, and Strong Institutions.

In this context, is there a remedy for assuming planetary citizenship? How can we continue to conceive of such a utopia in light of such events and occurrences? To what extent is it pertinent to rethink the discussion of constitutional states of law? To what extent are we prepared to construct the formation of a universal person who embodies the values, feelings, and moral criteria necessary to build such an archetype of human being? Our view is that it is feasible through the implementation of a new type of pedagogy, which we will discuss below.

3.3.2 Humanistic Education

The points established by the UN regarding the SDGs require, for their realization, the development of citizenships with an ecumenical perspective (*oikoumenikós* or *oecumenĭcus*), global, cosmopolitan, universal, relational, and planetary, which implies an analogical vision of the *polis*, where each human being practices solidarity, hospitality, and responsibility in caring for oneself and for the Other, as well as for nature and everything around us.

This will allow us to reclaim our civic circumstances, rescuing our human condition and our transformation into subjects. The analog citizen is a critic of efficiency in favor of dividends, that is, of the logic of interests over justice, of the market above the common good. The era of the 2030 Agenda and the SDGs is a ruthless and brutal period. The planetary citizen does not adapt to the violence of positivist normativity, rejects corruption, and always seeks subjective affirmation, the transcendence of the animal condition—in other words, the transformation into a critical subject. The fulfillment of the 2030 Agenda is not possible without an economic ethic capable of valuing commutative, productive, and distributive justice. At the same time, a philosophical ethic is indispensable, one that allows us to explore a previously unimagined path, illuminating the choices of the citizen, practicing virtuous citizenship, and avoiding decisions made as a dependent and alienated individual. Determination, planetarity, sanity and caution, autonomy, and community life—these are the virtues of the citizen and analog education. In contrast, the reified individual is based on corruption, blind obedience, submission to power and the state, monetary fetishism, violence, inequality, and lack of judgment.

Now then, what is the relationship between the SDGs and humanistic education? First, humanistic education involves focusing its efforts on the development of judgment in both learners and educators—that is, not only students but also teachers. In other words, to engage in praxis. It is to know and apply principles such as equality and others. It is to apply induction, deduction, abduction, materialist historization, and methodological dialectics. On this subject, the German hermeneutic philosopher Hans-Georg Gadamer notes:

> What constitutes the difference between the fool and the discerning person is that the former lacks the capacity for judgment; that is, he is not in a position to correctly subsume nor, consequently, to correctly apply what he has learned and what he knows. The introduction of the term "capacity for judgment" (*Urteilskraft*) in the eighteenth century thus attempts to adequately reproduce the concept of *judicium*, which must be considered a fundamental spiritual virtue. (Gadamer 1988, p. 61).

We see that to have judgment about something implies incorporating and absorbing the specific from a generality. That is, to insert the particular into the universal, to subsume the abstract into the concrete.

This is one of the tragedies of contemporary instruction: failing to take into account the capacity for judgment of both educators and learners, falling into positivist, constructivist, and functionalist theories that do not form citizens oriented toward the capacity for judgment. It is important to consider that, in order to receive a humanistic education, one must first have the opportunity to be educated; that is, one must have the means, infrastructure, objects, and subjects appropriate for the educational process and, subsequently, make them available to citizens.

Citizenship (*politeia*), which is also a form of government or constitution and state, protects the individual and enables the realization of democracy in the republican idea of Athenian substance. In this sense, there is an articulation between *paideia* or education and the citizen (*polites*). Thus, the *praxis* of democracy (*demokratía*) implies the existence of a planetary citizen and an education of the human being (*paideía toû anthrōpou*). In this way, we realize that, to avoid corruption and unsustainability, a humanistic education and citizen are needed. What will their characteristics be?

It will entail a critical and participatory citizenship, that is, taking into consideration the Aristotelian idea of the citizen:

> It is clear who the citizen is: whoever has the possibility of participating in deliberative or judicial functions, we call a citizen of that city; and we call a city, to put it briefly, the set of such citizens sufficient to live in self-sufficiency. (Aristotle 2008, pp. 155–156)

We see that the Stagirite envisions the citizen as an active agent in the deliberative and decision-making processes within their cities. The citizen is one who participates actively, constantly, and continuously in public affairs. For Aristotle, self-sufficiency has an anthropological meaning, since the self-sufficient citizen is the one who applies reason in their concrete praxis generated in the *polis*. That is, a citizen is a subject who subsumes the ecumenical into the personal.

At present, it is necessary to rethink a dynamic citizenship that defends the SDGs and even goes beyond them, proposing new pathways for collective improvement. This should not only formally address poverty reduction, peace, sustainable cities, and quality education, but also the equitable distribution of social wealth, the right to protest, free education, the real—not merely formal—right to security, protection of electronic data and free internet access, the human right to be accepted in any country as expressed in migration rights, political association, euthanasia, the right to a dignified death, desired motherhood, and, in general, the supreme right to a dignified life. All of this will be possible to the extent that we have both an analog citizenship and a humanistic education—neither of which have been fully realized in practice or sufficiently addressed in formal terms—that fulfill the SDG goals and aim to surpass them. Determination, prudence, and *becoming what we are* the virtues of the planetary citizen and participants in a humanistic education, which implies a new kind of sustainability.

In this way, we visualize the articulation between a planetary citizenship of an analogical nature, and a humanistic education based on the protection of subjective rights, support for science and technology, the promotion of cutting-edge research and the conscious practice of cordial and friendly reasoning above instrumental reasoning, so frequent in our environment.

3.4 Discussion

Regarding the comprehensive approach of our text, it is necessary to point out that the very idea of a deep understanding of the SDGs cannot be merely numerical and quantitative, since understanding allows us to extend a broader view of the matter than simply visualizing data, figures, and enumerations. It is linked to intentionality and not so much to extension, to subtlety above the machinic device, and to the hermeneutic virtues of ontological and epistemic origin over technocratic competences.

Regarding the gaps and shortcomings of the SDGs, it should be noted that the SDG objectives are seven years from their deadline and are far from being achieved. The officials themselves state:

> Progress on more than 50% of the SDG targets is weak and insufficient, and 30% are stalled or have regressed. These include essential targets on poverty, hunger, and climate. If we do not act now, the 2030 Agenda could become the epitaph for the world that could have been.
>
> [...] Funding to halt climate change also falls far short of commitments, and developed countries have not provided the $100 billion per year promised since 2020.
>
> [...] The SDGs are the universally agreed roadmap to overcome economic and geopolitical divisions, restore trust, and rebuild solidarity. The lack of progress means that inequalities will continue to deepen and the risk of a fragmented world operating at two speeds will increase. No country can afford the failure of the 2030 Agenda. (United Nations 2023, p. 2)

Regarding Humanistic Education, there are obstacles stemming from low government investment, privatization, limited civic awareness, war, lack of infrastructure in school environments, violence, absence of free education, lack of school meal programs, lack of union and association freedoms, insistence on univocal instructional models, total neglect of rural education, and digital illiteracy. These issues were exacerbated by the educational setbacks caused by the COVID-19 pandemic, where even in developed countries stagnation reached alarming levels. In developing countries, dropout and school abandonment rates, especially in rural areas, have reached catastrophic proportions. According to data from the Ministry of Public Education, during the 2020–2021 period—clearly attributable to the pandemic—school dropout rates in Mexico increased by 11.6 points in upper secondary education, 2.9 in secondary, and 0.5 in primary education (Children's Rights Network 2023).

Finally, regarding Planetary Citizenship, it is clear that the number of armed conflicts is a major barrier to progress toward this ideal. The populist triumph of the right in much of the world, and resistance to irregular migration from developed countries suffering under incompetent, corrupt, and despotic regimes, have created human flows that no wall or legislation can contain. The lack of a global agenda on this issue makes it difficult to imagine progress without a broad global consensus that makes governance an essential element for assessing the performance of small nation-states, which political leaders stubbornly seek to preserve. This is the area with the greatest lag in realizing the 2030 Agenda.

3.5 Conclusions

We have reflected on the role of citizenship and education in the establishment and achievement of the Sustainable Development Goals (*Sustainable Development Goals*), developed by United Nations bodies to address global challenges. These goals have been discussed in general terms, and in more detail, Goals 4 and 16, which pertain to education and peace. We have highlighted the need to design a genuine education aimed at preparing students and teachers, as well as citizens in general (both those who govern and those who are governed). In conclusion, we can say that without an education and citizenship of a planetary nature, and with the passive preservation of current social relations of production, it will be impossible to achieve all the objectives mentioned. Although our review found significant gaps in Humanistic Education and Planetary Citizenship, we do not wish to be pessimistic. The major differences in development are also linked to the evolution and strength of institutions that enable a state to defend itself against global corporations with budgets and resources exceeding the GDP of most countries, and against abusive, deceitful, and demagogic leaders who, unfortunately in recent years, have held power in the world's most powerful countries, such as the United States, Russia, and China, and who have, in one way or another, hindered progress on the 2030 Agenda.

The strengthening of autonomous oversight bodies that monitor, supervise and limit the exercise of power by despots, who from time to time seize power, is one of the fundamental elements for making steady progress toward a planetary citizenship and a humanistic education that enables us to care for our beautiful and fragile planet.

Acknowledgements Napoleón Rosario Conde Gaxiola acknowledges Instituto Politécnico Nacional and Secretaría de Investigación y Posgrado (SIP) for the support provided for the good development of the Project 20250770.

Angel Eustorgio Rivera González acknowledges the support provided by the Instituto Politécnico Nacional through the Researcher Performance Incentive Program (EDI) and the Secretaría de Investigación y Posgrado (SIP) for the support provided for the good development of the Project 20253662, as well as the SNII of SECIHTI for the grant awarded.

References

Aristotle (2008) Politics. Gredos

Children's Rights Network (2023) School dropout rate in Mexico. https://blog.derechosinfancia.org.mx/2022/10/05/abandono-escolar-de-ninas-ninos-y-adolescentes-en-mexico/

Conill Sancho J (2009) Poverty reduction and the promotion of freedom from the perspective of discourse ethics. In: Cortina A, Pereira G (eds) Poverty and freedom. Eradicating poverty from Amartya Sen's capabilities approach. Tecnos, pp 151–162

Cortina A (2009) Poverty as a lack of freedom. In: Cortina A, Pereira G (eds) Poverty and freedom. Eradicating poverty from Amartya Sen's capabilities approach. Tecnos, pp 15–30

Daly H (1996) Beyond growth: the economics of sustainable development. Beacon Press

Dohmen T, Falk A, Huffman D, Sunde U (2009) Homo reciprocans: survey evidence on behavioral outcomes (O. U. Press, Ed.). Econ J 119(536):592–612. https://www.jstor.org/stable/20485333

El País (2023) War between Israel and Gaza: summary of 12/28/2023. El País. https://elpais.com/internacional/2023-12-28/guerra-entre-israel-y-gaza-en-directo.html

EL PAÍS Channel (2023) Ukraine War I Nearly 10,000 people have died in the armed conflict with Russia. [Video file]. Youtube. https://www.youtube.com/watch?v=gbOUq5PVC0o

Fehr E, Gächter S (1998) Reciprocity and economics: the economic implications of homo reciprocans. Eur Econ Rev 42(3–5):845–859. https://www.sciencedirect.com/science/article/abs/pii/S0014292197001311

Gadamer H (1988) Truth and method. Foundations of a philosophical hermeneutics. Ediciones Sígueme

Gil C (2018) Sustainable development goals. Papers on social-ecological relations and global change, vol 140(1), pp 107–118

Hernández Orozco F (2023) Which countries are at war in 2023? Expansión. https://expansion.mx/mundo/2023/10/20/que-paises-estan-en-guerra-2023

Jackson T (2006) The Earthscan reader in sustainable consumption. Earthscan

Kroll C (2015) Sustainable development goals: are the rich countries ready? Sustainable governance indicators—Bertelsmann. Bertelsmann Stiftung

Leff E (1988) Environmental knowledge: sustainability, rationality, complexity, power. Siglo XXI/UNAM/UNEP

Leff E (2019) The fire of life. Heidegger and the environmental question. Siglo XXI

Morin E (2006) Flat earth. Nueva Visión

Novo M (2006) Sustainable development. Its environmental and educational dimension. Pearson/UNESCO, Madrid

Nussbaum MC, Sen A (1996) The quality of life. FCE

Pearce DW, Atkinson G (1992) Are national economies sustainable?—Measuring sustainable development. CSERGE GEC Working Paper 92-II. University College London

Rivas-Tovar LA (2021) Strategic management. Trillas

Rivas-Tovar LA (2023) APA7 standards. Edition, structure, citations and references. https://www.researchgate.net/publication/357046089_NORMAS_APA_7_EDICION_ESTRUCTURA_CITAS_Y_REFERENCIAS

Rockström J, Steffen W, Noone K, Persson Å III, FS, Lambin E et al (2009) Planetary boundaries: exploring the safe operating space for humanity. Ecol Soc 14(2):1–34. https://www.jstor.org/stable/26268316?seq=1

Sarvajayakesavalu (2015) Addressing challenges of developing countries in implementing five priorities for sustainable development goals. Ecosyst Health Sustain 1(24):24

Sen A (2010) The idea of justice. Taurus

Simon J (1983) The ultimate resource. Princeton. https://www.degruyter.com/document/doi/10.1515/9780691214764/html

Turner RK (1993) Sustainability: principles and practice. In: Kerry Turner R (ed) Sustainable environmental economics and management: principles and practice. Belhaven Press

United Nations (UN) (2023) The sustainable development goals report 2023: special edition. A rescue plan for people and planet. UN. https://unstats.un.org/sdgs/report/2023/The-Sustainable-Development-Goals-Report-2023_English.pdf

Weitz Å, Nilsson M, Tengrgren S (2015) Sustainable development goals for Sweden: insights for a national agenda. Working Paper No. 25-10: Stockholm Environment Institute

Beyond the SDGs: Degrowth as an Alternative

4

Gibrán Rivera González, Angel Eustorgio Rivera González, Francisco Javier Carrillo, and Aydé Cadena López

Abstract

The COVID-19 pandemic revealed not only the scale of global challenges but also the profound inequalities and shortcomings in our collective response to crises of this magnitude. This situation calls for reflection on the functioning of the prevailing economic system: capitalism, which promotes unchecked growth, the use of fossil fuels, globalization, and the unsustainable exploitation of nature. This chapter explores how degrowth, as an alternative to the capitalist system, can contribute to achieving the Sustainable Development Goals (SDGs), seeking a comprehensive approach to balance environmental, social, and economic sustainability on the planet, improve quality of life, generate social welfare, establish more respectful relationships with nature, and foster harmony among nations. It thus underscores the need to rethink the current economic paradigm in search of more sustainable and equitable solutions.

Keywords

Degrowth · Capitalism · SDGs · Global challenges

4.1 Introduction

We are currently living in a context characterized by rampant economic growth, accompanied by excessive accumulation and concentration of wealth in the hands of a minority (ECLAC 2016). This condition has been driven, among other phenomena, by globalization and unprecedented technological advances (De Benoist 1996). However, this approach of perpetual growth has caused and fueled a series of increasingly frequent and severe problems and crises affecting various areas of society (Bisk 2012; The Lancet Planetary Health 2024). This situation presents major challenges for the sustainability of the planet and human well-being. One such challenge is to develop a new, or at least a more complex, understanding between the social sphere and the physical reality in which we live (Spash 2017).

Globalization, for example, has fostered an interconnectedness among countries like never before, resulting in a high dependence on global supply chains, encouraging excessive

G. R. González (✉) · A. E. R. González · A. C. López
Unidad Profesional Interdisciplinaria de Ingeniería y Ciencias Sociales y Administrativas del Instituto Politécnico Nacional, Ciudad de México, México
e-mail: griverag@ipn.mx

A. E. R. González
e-mail: aerivera@ipn.mx

A. C. López
e-mail: aydecadena@gmail.com

F. J. Carrillo
World Capital Institute, Secretariat Office, Jurica, Querétaro, QRO, México
e-mail: fjcarrillo@worldcapitalinstitute.org

P. Pérez-Hernandez and R. Huerta Cuervo (eds.), *Social Well-Being and the Challenges in the Sustainable Development Goals*, Sustainable Development Goals Series,
https://doi.org/10.1007/978-3-032-14294-8_3

resource exploitation (Bello 2006), and as we recently experienced during the pandemic, creating systemic vulnerability to global health crises (Sigler et al. 2021; Lupu and Tiganasu 2022). Technological development, in turn, has led to issues such as job losses due to automation and has increased inequalities, clearly reflected in the digital divide (Lima et al. 2021). Artificial intelligence is now seen as a recent threat that is expected to exacerbate these challenges (Federspiel et al. 2023).

In response to these problems, various initiatives have emerged aiming to address global challenges in a comprehensive manner. The Sustainable Development Goals (SDGs), proposed by the United Nations Development Programme, are one such initiative. While these goals are based on the recognition of global issues such as poverty, hunger, inequality, and climate change, among others, the idea that economic growth is fundamental to human well-being remains at the core of this initiative. However, as we argue in this work, although the SDGs seek to address some of the problems currently facing humanity, they simultaneously undermine the resolution of these global issues by continuing to promote economic growth (SDG 8) as a central pillar of the proposal. Without losing sight of this fundamental concession made to keep economic growth as one of the SDGs, we also show that the very structure of the current capitalist system hinders the effective achievement of the remaining SDGs.

Upon reflection, it becomes clear that the logic of constant economic growth, excessive dependence on fossil fuels, unsustainable exploitation of natural resources, perpetuation of inequalities, pursuit of short-term profits, and the externalization of environmental and social costs in current production systems are practices deeply rooted in capitalism that have contributed to the emergence and exacerbation of many of the problems the SDGs aim to solve. This situation suggests that a change "within the system" is not sufficient to address today's global crises; what is needed is a change "of the system."

Based on this reasoning, we propose the need to consider radical and transformative alternatives that contribute to achieving the SDGs, while not overlooking, as previously mentioned, the exclusion of SDG 8, which reflects the intention to continue promoting economic growth. To this end, we suggest that degrowth (Schmelzer et al. 2022) offers an alternative that goes beyond mere adjustments to the prevailing system. Degrowth advocates for a democratic transition toward a more equitable and sustainable society, challenging the logic of perpetual economic growth by recognizing that we live on a planet with finite resources and clearly identified planetary boundaries (Rockström et al. 2009).

In this regard, we develop the idea that the principles of degrowth are closely related to several of the SDGs. For example, Goal 1, ending poverty, is directly linked to wealth redistribution, a fundamental principle of degrowth. Similarly, Goal 12, responsible consumption and production, resonates with degrowth, which advocates for the promotion of more sustainable and conscious production and consumption practices, thus challenging the logic of unlimited economic expansion promoted by capitalism.

To develop these ideas, the chapter is organized as follows. First, we describe a series of capitalist practices and principles and argue how these contradict the SDGs. Second, we examine the origin and intent of the SDGs as an initiative to address several current global problems, while acknowledging their emphasis on maintaining SDG 8 Economic Growth as a central part of the proposal. Third, we discuss the alternative model of degrowth and its contribution to achieving the SDGs. Finally, we present our concluding thoughts and invite a deeper reflection to improve our collective actions on the planet.

4.2 The Capitalist System and the (Non-)fulfillment of the SDGs

Three scenarios are described below that illustrate some fundamental interactions and situations that reinforce the reproduction and perpetuation of current capitalist practices, as well as the deterioration of living conditions for people and the destruction of the planet and biodiversity, and therefore, their inability to contribute to the achievement of the SDGs. The interactions presented are not exhaustive; rather, the existence of many other interactions within and between scenarios is acknowledged. The purpose of this simplification is to illustrate how the reproduction of capitalist practices affects at least three fundamental dimensions of people and the planet: environmental, social, and productive-economic.

4.2.1 Social Dimension

The accumulation and concentration of wealth by a minority are essential characteristics of a capitalist system (Rehbein 2020) since both often translate into greater and better opportunities to generate more wealth. Thus, countries, companies, and individuals with greater economic resources can access more profitable financial investments and services, as well as privileged business opportunities (Williams 2016). As we will discuss below, this capitalist dynamic not only generates economic inequalities but also social ones, thereby undermining efforts to reduce poverty (SDG 1), inequality (SDG 10), and hunger (SDG 2).

The accumulation of wealth, understood as one of the main objectives in a capitalist system, contributes to the perpetuation of inequality between the rich and the poor (SDG 10) (Williams 2016). The case of income earned through labor in a capitalist system clearly demonstrates how economic inequalities generated by exponentially different income levels exacerbate disparities in other areas (SDG 10). Those with access to fewer economic resources are often affected by limited opportunities to access basic services such as education (SDG 4), social security, health, and healthy lifestyles (SDG 5). In both health and education, it is a common capitalist practice for these industries to prioritize profitability over quality health and education, thereby compromising the integrity of these essential rights as public goods (Sell and Williams 2020). Moreover, to this day, countries where capitalism prevails have shown a dramatic decline in human well-being (Sullivan and Hickel 2023), with a large portion of their populations living in conditions of extreme poverty (SDG 1) and malnutrition (SDG 2).

Also related to the concentration of economic power in small segments of the population are problems that undermine social cohesion and contribute to institutional corruption, thus threatening the principles of peace, justice, and strong institutions (SDG 16) (Laurell 2015). This is observable in various countries where the power held by large corporations influences government decisions in favor of transnational companies, thereby prioritizing corporate interests over the public good (Royo-Bordonada 2019). In the long term, the perpetuation of such practices leads to the weakening of systems such as health, social security (SDG 5), and education (SDG 4) (Laurell 2015), a situation that becomes especially evident in times of crisis, such as the recent COVID-19 pandemic.

Another aspect that affects the social dimension is the competition promoted in capitalist systems; this is particularly evident in the labor market. Since an open market—a necessary condition in a capitalist system—encourages competition among different productive actors, they sometimes resort to cost-cutting measures that can lead to labor exploitation, resulting in unfair labor practices and precarious working conditions (Janardhan 1997) which undermines decent work (SDG 8) and contributes to the perpetuation of poverty (SDG 1).

4.2.2 Environmental Dimension

The constant aspiration of economic growth as a pillar of capitalism tends to drive an unsustainable model of consumption and production (SDG 12) (Ruuska 2017). Over time, this model generates significant environmental impacts, largely due to the indiscriminate use of fossil fuels (SDG 6), accelerating climate change (SDG 13), degrading marine (SDG 14) and terrestrial ecosystems (SDG 15), and causing excessive pollution of soil, air, and water (SDG 6) (Spash 2017).

Thus, in its pursuit of resources for excessive and unsustainable production and consumption (Schor 2005), the capitalist system perpetuates unsustainable practices of natural resource extraction (Foster and Clark 2009), which in turn undermines food sovereignty (SDG 2), access to clean water (SDG 6), and clean energy (SDG 7). Furthermore, the heavy reliance of many capitalist systems on the excessive use of fossil fuels exacerbates the challenge of providing affordable and sustainable energy (SDG 7) (Altvater 2007), while simultaneously accelerating climate change (SDG 13) (Soeder 2021).

Additionally, the deterioration of global ecosystems has significantly accelerated the emergence and spread of infectious diseases such as that caused by SARS-CoV-2 (Pfenning-Butterworth et al. 2024). As well documented in the literature, the aforementioned environmental degradation, leads to new and more frequent interactions among people, domestic animals, and wildlife, thereby increasing human contact with pathogenic organisms present in species with which humans have close and regular interaction (Daszak et al. 2001; Patz et al. 2004; Nava et al. 2017; Gibb et al. 2020). This, in turn, directly or indirectly promotes the emergence and spread of zoonotic infectious diseases (Patz et al. 2004; Gibb et al. 2020; Destoumieux-Garzón et al. 2022), which currently account for more than 75% of emerging infectious diseases worldwide associated with human contact with wildlife and domestic animals (Patz et al. 2004). The situation becomes alarming when it is recognized that the emergence of these diseases—whose pathogenic agents are transmissible between animals and from animals to humans, such as HIV, H1N1, Ebola, Nipah, Zika, Chikungunya, and COVID-19—has increased significantly in recent times and will become increasingly common (Jones et al. 2008; Destoumieux-Garzón et al. 2022).

The expansion of urban environments is a clear example of this situation. This expansion is accompanied by the construction of roads, infrastructure, and changes in land use. These factors have significantly contributed to forest encroachment and, consequently, excessive deforestation (Zipperer et al. 2012; Pfenning-Butterworth et al. 2024). The consequences of these practices include the destruction of natural habitats, the resulting loss of biodiversity (SDG 14 and SDG 15), and new interactions between humans and other species with the potential to generate and spread emerging infectious diseases (Lindahl and Grace 2015).

4.2.3 Productive-Economic Dimension

Perhaps the most widely analyzed dimension in the literature related to the strategies and productive systems of the capitalist system is the productive-economic dimension. The capitalist system promotes the generation of (economic) wealth through the sale, distribution, and consumption of goods and services in the market. There are agents who produce these goods and others who require them, thus generating cycles of value exchange that benefit both parties. In the economic cycle, the system of production and consumption is one of the most important elements of capitalism and, in this sense, the following reflections are worth considering, as their answers call into question the viability of the capitalist system: (1) Is it possible for companies, the economic units that produce goods, to use the inputs they require without limit?; (2) Are raw materials unlimited resources that will always exist?; (3) Is it necessary to continuously and massively increase the consumption of goods that are not essential for human survival?; (4) Can there be another mechanism besides economic-financial value that motivates entrepreneurs to improve citizens' lives?; (5) Is it fair and humane for citizens of a certain country to enjoy the benefits of the work and commitment of other citizens who live and work in subhuman conditions?

Reflecting on these questions allows us to identify the most significant and unjust contradictions of a system that revolves around the rationality of *homo economicus* and, on the other hand, almost entirely neglects and ignores *homo reciprocans* (Fehr and Gächter 1998, 2000; Dohmen et al. 2009), which is also part of our nature. Eliminating poverty (SDG 1) and reducing inequalities (SDG 10), for example, requires a strong commitment to distributing more fairly and equitably the resources obtained from the sale of goods and services in the market (Bennett and Grabs 2024). Generating decent work and economic growth (SDG 8) means that those who dominate the market (shareholders and business owners) must be willing to earn a little less in order to invest more and better in working conditions and people's safety. Responsible production and consumption (SDG 12) is meaningful in a world where companies and their decision-makers consciously promote the elimination of single-use economic thinking among their clients and consumers (Chang and Hung 2023). We must rethink and reconfigure our conception of value and consider much more comprehensive proposals (Carrillo 2002) that allow us to understand, define, value, and distribute what is important in a more holistic, just, and equitable manner.

The system in which we live has eroded our best human qualities, allowing us (with our limited rationality) to normalize (wrongly) indifference and disregard for those who live beside us, arrogance, and contempt for collaboration, giving way—relentlessly and without restraint—to the fierce competition that dominates the free market, where everything is for sale, even human dignity.

As shown in the previous section, the evidence is overwhelming in demonstrating that the reproduction and perpetuation of these interests, practices, and structures of the current dominant system have dangerously exacerbated many of the global problems we currently face across the planet (Martin 2021). We live in times of crisis: climate change, ecosystem collapse, biodiversity loss, natural disasters, civil wars, extreme poverty, hunger, precarious employment, pandemics, and economic and financial crises (Spash 2017).

In light of these problems and the urgent need to address them, we describe below the initiative of the United Nations Development Programme, which identifies the so-called Sustainable Development Goals (SDGs), aimed at balancing environmental, social, and economic sustainability on the planet. We also explicitly question the inclusion of SDG 8, which seeks to maintain the notion of economic growth as the central axis for development and sustainability.

4.3 Origin, Contribution, and Challenges of the SDGs in the Face of Global Problems

The Sustainable Development Goals (SDGs), also known as Global Goals, were established in 2015 by the United Nations as part of the 2030 Agenda for Sustainable Development, seeking to address the impact of the multiple interconnected crises we currently face as humanity (ECLAC 2016). Collectively, they aim to tackle global challenges such as poverty, hunger, violent conflict, inequality, human rights abuses, gender inequality, environmental degradation, climate change, among others (ECLAC 2016). The aim is for all 17 of these goals, with their 169 targets and 231 indicators, to be achieved by 2030 (ECLAC 2016).

According to ECLAC, three main characteristics of the SDGs can be highlighted: they are universal, transformative, and civilizational (ECLAC 2016). They are universal because they address global challenges that require the collective effort of all to achieve. They are transformative because they offer a vision that encompasses people, the planet, prosperity, peace, and partnerships, integrating economic, social, and environmental dimensions. Finally, they are civilizational because they seek equality and non-discrimination, always aiming to respect, protect, and promote human rights.

Although resources, efforts, and international cooperation have been mobilized to achieve the SDGs, today, a little over halfway to the 2030 Agenda deadline, the results are discouraging due to inadequate and insufficient capacity and methodology for their implementation (Spangenberg 2017). These results can largely be explained by the fact that, while countries aim to advance the SDGs, they simultaneously perpetuate capitalist system practices, using GDP growth as the main indicator of national development (Schmelzer et al. 2022). This situation is also reflected in the formulation of SDG 8, which prioritizes economic growth as the driving force of development. In addition, other phenomena such as insufficient funding, global conflicts and political instability, ongoing pandemics and global crises, and the lack of citizen participation continue to pose current challenges to achieving the SDGs (Kempf and Hujo 2022).

Taking into account the points discussed in the previous and current sections, Table 4.1 shows how various capitalist practices undermine the achievement of the SDGs.

4.4 Degrowth as an Alternative for Achieving the SDGs

Once it is recognized that the human species has become a genuine geological force, critically altering the cycles of the biosphere through mechanization, productivism, and the technology enabled by fossil fuels (Schmelzer et al. 2022), it is relevant to discuss degrowth. The term originated in the 1970s and 1980s with Nicholas Georgescu-Roegen's thermodynamic theory of economic processes. However, it only acquired political, economic, and social significance in the 2000s, when the effects of the neoliberal system became evident (Kallis et al. 2018).

On the one hand, degrowth is considered a political project that simultaneously critiques the dominant system and proposes a vision for the future (Schmelzer et al. 2022), with the potential to keep humanity within a safe operating space (Kronenberg et al. 2024). Thus, degrowth promotes a set of common principles for what that future could look like, policies for how to get there, and strategies to achieve such a transformation. As Stefania Barca explains, referring to degrowth as an "*umbrella*" term, not only because it encompasses diverse perspectives but also because it seeks to "protect":

> In a world constantly battered by the heavy rains of ecological degradation, impoverishment, and austerity measures, as well as by the relentless heat of overconsumption, overproduction, and the financialization of everything, opening an umbrella and creating a space for alternatives to converge and discuss (or practice) what they seek is exactly what the term degrowth has been doing over the past decade, especially in Europe (Stefania-Barca 2018:379).

Degrowth is also understood as an economic movement that challenges the principles of orthodox economics and, in contrast, seeks to go beyond abstract theories or universal models (Kallis et al. 2015). It fundamentally questions the principles of growth economics and price-setting, such as free choice, competition, and efficiency (Windegger and Spash 2022). The main critique is that the market it is not the harmonious space where exchange takes place under conditions of equality; on the contrary, all the failures it has exhibited, such as monopolies and opportunistic behavior, have resulted in the vast majority of the population being poor and the planet Earth being destroyed at a rapid pace (Spash 2021).

In this regard, Leonhardt (2014) asserts that degrowth represents a fundamental critique, especially of the principle of economic efficiency, as it prompts reflection on efficiency *in what, for what, or for whom.* According to this author, degrowth highlights the fact that markets have not been able to provide better lifestyles with current modes of production and consumption, but have instead generated a rebound effect of poverty, pollution, and crisis.

In light of this situation, Alenda-Demoutiez (2022) notes that approaches such as degrowth become necessary, as they seek alternative pathways to the principle of economic efficiency and industrialization. This is achieved through alternative forms of exchange that take into account the historical and contextual development of different regions, and not only those of European or North American countries.

On the other hand, Spash et al. (2023) argue that degrowth can also be understood as a social movement, as it is grounded in an activist stance that invites us to *think outside the box.* For the author, more than a scientific paradigm, it represents a challenge to the current way of life and to the very worldview of humanity, as it seeks to understand the dialectical relationship between human beings and Nature; a new relationship in which humans are no longer seen solely as maxi-

Table 4.1 SDGs and capitalist practices and principles

	Sustainable development goal	What is the goal?	Capitalist practices and interests
1	No poverty	End poverty in all its forms everywhere by 2030	Accumulation and concentration of wealth among minorities
2	Zero hunger	End hunger, achieve food security and improved nutrition, and promote sustainable agriculture	Overexploitation and depletion of natural resources
3	Good health and well-being	Ensure healthy lives and promote well-being for all at all ages	Accumulation and concentration of wealth among minorities
4	Quality education	Ensure inclusive and equitable quality education and promote lifelong learning opportunities for all	Accumulation and concentration of wealth among minorities
5	Gender equality	Achieve gender equality and empower all women and girls	Lack of inclusion and participation
6	Clean water and sanitation	Ensure availability and sustainable management of water and sanitation for all	Overexploitation and depletion of natural resources
7	Affordable and clean energy	Ensure access to affordable, reliable, sustainable, and modern energy for all	High dependence on fossil fuels
8	Decent work and economic growth	Promote sustained, inclusive, and sustainable economic growth, full and productive employment, and decent work for all	Unlimited economic growth Labor exploitation
9	Industry, innovation and infrastructure	Build resilient infrastructure, promote inclusive and sustainable industrialization and foster innovation	Overexploitation and depletion of natural resources
10	Reduce inequalities	Reduce inequality within and among countries	Wealth accumulation Inequalities
11	Sustainable cities and communities	Make cities and human settlements inclusive, safe, resilient, and sustainable	Unlimited economic growth Globalization
12	Responsible consumption and production	Ensure sustainable consumption and production patterns	Single-use economy
13	Climate action	Take urgent action to combat climate change and its impacts	High dependence on fossil fuels
14	Life below water	Conserve and sustainably use the oceans, seas, and marine resources for sustainable development	Exploitation and depletion of natural resources
15	Life on land	Sustainably manage forests, combat desertification, halt and reverse land degradation, and halt biodiversity loss	Exploitation and depletion of natural resources Species extinction
16	Peace, justice and strong institutions	Promote peaceful and inclusive societies for sustainable development, provide access to justice for all, and build effective, accountable, and inclusive institutions at all levels	Privileges for minorities
17	Partnerships for the goals	Revitalize the global partnership for sustainable development	Lack of inclusion and participation

Source Own elaboration

mizers and Nature is no longer conceived merely as a provider of natural resources.

Furthermore, degrowth as a social movement implies reconsidering the relationships established among human beings, recognizing and addressing the interconnectedness of ecological, economic, and social issues. For this reason, Spash and Guisan (2021) argue that degrowth helps to integrate social issues that have been neglected by conventional economics, such as discrimination, exclusion, forced migration, public health, education, among others.

In this way, degrowth has developed as a normative concept whose applications can be analytical or practical, and whose stance seeks to offer an alternative to the principles of orthodox economics, asserting that ecological, economic, and social sustainability require a radical reorganization of reality as a whole (Kallis et al. 2018).

The criticisms of degrowth cannot be overlooked, especially those of Kate Raworth in her *doughnut* theory (Spash et al. 2023) and of other thinkers who claim that abandoning orthodox economic growth is impossible or that degrowth proposals are, to some extent, utopian illusions, as human and contextual conditions make their implementation impossible (Kallis 2018).

However, it cannot be denied that the radical stance of degrowth, along with other alternative approaches such as ecofeminism (Howell 1997; Warren 2000), ecosocialism (Löwy 2005; Kovel 2008), and ecological social economics (Spash 2017), have envisioned the possibility of a different path. According to Kallis (2018), while degrowth does not completely eliminate the paradigm of orthodox economics, it does conceive of alternative approaches that no longer see it as the only possible way forward.

Thus, the analytical and practical scope of degrowth makes it relevant in the context of achieving the SDGs, as it offers a comprehensive vision that recognizes the interconnectedness and interdependence of the issues addressed by the 2030 Agenda. One reason for the discouraging progress of this agenda is that each issue has been considered in isolation, and its treatment has been reduced to the economic logic of efficiency. In this context, degrowth can help to rethink the SDGs based on different practices and interests that go beyond this logic. Table 4.2 presents a proposal on how each SDG can be considered from the perspective of degrowth practices and interests.

4.5 Conclusions

We concur with the perspective expressed by The Lancet Planetary Health (2024), which recognizes that there will always be positions of power and influence under the status quo that resist changing the current system. However, we are convinced that embracing new forms of interaction among humans and with all else requires abandoning deeply rooted practices that have led us to believe that market dynamics, competition, and unrestricted economic growth are the best options for interacting with one another and with our environment. From our perspective, degrowth entails rethinking our priorities and our value frameworks. Degrowth means acknowledging that not everything revolves around the market, that not everything is about supply and demand.

As the authors of this document, we are aware that we cannot detach ourselves from our acquired capitalist responsibility, but we also believe that we can and must critically identify and acknowledge our mistakes in order to change our behaviors and attempt to reduce the harm caused by our decisions. We also recognize that not everyone thinks alike and that there will be capitalists who cling to what they have and what they sell. We hope that prudence prevails among us all and that the necessary reflection takes place to foster better understanding and to live with justice, equity, and harmony with others—those who are part of us—and with the planet we all share. We must coexist without destruction.

To achieve this, one of the most important elements for changing the way people think and, consequently, the way they act toward others is education. In this regard, it is essential that we conceive and develop initiatives, strategies, and mechanisms that design, promote, and disseminate an education grounded in communal responsibility and collective consciousness, capable of inspiring new ways of thinking about our social

Table 4.2 SDGs and degrowth

	Sustainable development goal	Degrowth practices and interests	Capitalist practices and interests
1	No poverty	More equitable redistribution of resources Reduction of the gap between rich and poor	Accumulation and concentration of wealth among minorities
2	Zero hunger	More equitable redistribution of resources Reduction of the gap between rich and poor	Overexploitation and depletion of natural resources
3	Health and Well-being	Decreased excessive production and consumption contributes to better water and air quality Prioritizing quality of life, equity, and environmental sustainability over economic growth	Accumulation and concentration of wealth among minorities
4	Quality education	Promotion of alternative educational models that enable humans to coexist with other species and with the planet as a whole	Accumulation and concentration of wealth among minorities
5	Gender equality	Social and economic justice for all, regardless of gender	Lack of inclusion and participation
6	Clean water and sanitation	Decreased excessive production and consumption contributes to better sustainable management of water resources	Overexploitation and depletion of natural resources
7	Affordable and clean energy	Decreased excessive production and consumption leads to lower use of non-renewable energy sources	High dependence on fossil fuels
8	Decent work and economic growth	Promote work and collective responsibility that replaces individual interest, in order to generate human well-being and not just economic development	Unlimited economic growth Labor exploitation
9	Industry, innovation, and infrastructure	Circular economy	Overexploitation and depletion of natural resources
10	Reduced inequalities	Sharing resources, skills, and knowledge to equalize conditions and human well-being	Inequalities
11	Sustainable cities and communities	Circular economy	Unlimited economic growth Globalization
12	Responsible consumption and production	Reduction of unnecessary consumption and resource waste	Single-use economy
13	Climate action	Reduction of greenhouse gas emissions by curbing economic growth	High dependence on fossil fuels
14	Life below water	Reduce the overexploitation of natural resources	Exploitation and depletion of natural resources
15	Life on land	Reduce the overexploitation of natural resources	Exploitation and depletion of natural resources Species extinction
16	Peace, justice, and strong institutions	Develop more inclusive and supportive organizational structures capable of making decisions for the majority	Privileges for minorities
17	Partnerships for the goals	Increase relational capital among individuals and societies as a foundation for greater collective awareness toward achieving planetary goals	Lack of inclusion and participation

Source Author's own work

and human actions. The outdated system that reinforces mercantilist and consumerist logic must be challenged through critical and collective thought, and we must make way for a logic that is much closer to collective responsibility and responsible, solidarity-based consumption.

There will be no social evolution without education that is conscious of the bad habits and individualistic practices acquired in and promoted by the capitalist system. Whether we are few or many, those of us who are convinced must strive, from our own positions, to practice citizenship that cares for other living beings on the planet, for the resources provided by the complex system we inhabit, and for the system itself, which we have pushed to dangerously inadequate limits for ensuring the well-being and prosperity of future generations. We humans are anthropologists before we are economists.

At the end of our lives, we will answer for our actions to the planet and to our descendants, and if, by then, our actions remain as harmful and unconscious as they have been so far—full of self-interest and arrogance—let us hope there will be no descendants left to mourn and endure the misfortunes caused by our dehumanized decisions.

Acknowledgements Gibrán Rivera González acknowledges the support provided by the Instituto Politécnico Nacional through the Researcher Performance Incentive Program (EDI) and Project 20254004, as well as the SNII of SECIHTI for the grant awarded.

Angel Eustorgio Rivera González acknowledges the support provided by the Instituto Politécnico Nacional through the Researcher Performance Incentive Program (EDI) and Project 20253662, as well as the SNII of SECIHTI for the grant awarded.

References

Alenda-Demoutiez J (2022) From economic growth to the human: reviewing the history of development visions over time and moving forward. Third World Q 43(5):1038–1055

Altvater E (2007) The social and natural environment of fossil capitalism. Soc Reg 43:37–59

Bello W (2006) The capitalist conjuncture: over-accumulation, financial crises, and the retreat from globalisation. Third World Q 27(8):1345–1367

Bennett EA, Grabs J (2024) How can sustainable business models distribute value more equitably in global value chains? Introducing "value chain profit sharing" as an emerging alternative to fair trade, direct trade, or solidarity trade. Bus Ethics Environ Respons 00:1–21

De Benoist A (1996) Confronting globalization. TELOS-ST LOUIS MO THEN NEW YORK-, pp 117–138

Bisk T (2012) No limits to growth. World Futures Rev 4(1):12–27. https://doi.org/10.1177/194675671200400104

Carrillo FJ (2002) Capital systems: implications for a global knowledge agenda. J Knowl Manag 6(4):379–399

Chang CT, Hung MF (2023) Mind the gap: analyzing factors associated with consumers' single-use product reduction. Sustain Prod Consum 36:75–87

Daszak P, Cunningham AA, Hyatt AD (2001) Anthropogenic environmental change and the emergence of infectious diseases in wildlife. Acta Trop 78:103–116

Destoumieux-Garzón D et al (2022) Getting out of crises: environmental, social-ecological and evolutionary research is needed to avoid future risks of pandemics. Environ Int 158:106915

Dohmen T, Falk A, Huffman D, Sunde U (2009) Homo reciprocans: survey evidence on behavioral outcomes. Econ J 119:592–612

ECLAC (2016) Taxation for inclusive growth, ECLAC, OXFAM, United Nations

ECLAC U.N. (2016) The 2030 agenda and the sustainable development goals: an opportunity for Latin America and the Caribbean

Federspiel F, Mitchell R, Asokan A, Umana C, McCoy D (2023) Threats by artificial intelligence to human health and human existence. BMJ Glob Health 8(e010435). https://doi.org/10.1136/bmjgh-2022-010435

Fehr E, Gächter S (1998) Reciprocity and economics: the economic implications of homo reciprocans. Eur Econ Rev 42:845–859

Fehr E, Gächter S (2000) Fairness and retaliation: the economics of reciprocity. J Econ Perspect 14(3):159–181

Foster JB, Clark B (2009) The paradox of wealth: capitalism and ecological destruction. Mon Rev 61(6):1–18

Gibb R et al (2020) Zoonotic host diversity increases in human-dominated ecosystems. Nature 584:398–402

Howell NR (1997) Ecofeminism: what one needs to know. Zygon 32(2):231–241

Janardhan V (1997) Globalization of capital, multinational corporations and labor: towards a perspective. Econ Polit Wkly L2–L9

Jones KE et al (2008) Global trends in emerging infectious diseases. Nature 451(7181):990–993

Kallis G, Kostakis V, Lange S, Muraca B, Paulson S, Schmelzer M (2018) Research on degrowth. Annu Rev Environ Resour 43(1):291–316. https://doi.org/10.1146/annurev-environ-102017-025941

Kallis G, Demaria F, D'Alisa G (2015) Introduction: degrowth. In: D´Alisa G, Demaria F, Kallis G (eds)

Degrowth. A vocabulary for a new era. Routledge, pp 1–17
Kempf I, Hujo K (2022) Why recent crises and SDG implementation demand a new eco-social contract. In: Financial crises, poverty and environmental sustainability: challenges in the context of the SDGs and Covid-19 recovery, pp 171–186
Kovel J (2008) Ecosocialism, global justice, and climate change. Cap Nat Soc 19(2):4–14
Kronenberg J, Andersson E, Elmqvist T, Łaszkiewicz E, Xue J, Khmara Y (2024) Cities, planetary boundaries, and degrowth. Lancet Planet Health 8(4):e234–e241
Laurell AC (2015) Three decades of neoliberalism in Mexico: the destruction of society. Int J Health Serv 45(2):246–264. https://doi.org/10.1177/0020731414568507
Leonhardt H (2014) Efficiency reconsidered: a social ecological economics approach. http://www.heterodoxnews.com/HEN/attach/hen173/Leonhardt2014a.pdf
Lima Y, Barbosa CE, dos Santos HS, de Souza JM (2021) Understanding technological unemployment: a review of causes, consequences, and solutions. Societies 11(50):2–17. https://doi.org/10.3390/soc11020050
Lindahl JF, Grace D (2015) The consequences of human actions on risks for infectious diseases: a review. Infection Ecol Epidemiol 5(1):1–11. https://doi.org/10.3402/iee.v5.30048
Löwy M (2005) What is ecosocialism? Cap Nat Soc 16(2):15–24. https://doi.org/10.1080/10455750500108237
Lupu D, Tiganasu R (2022) The implications of globalization on COVID-19 vaccination in Europe. Sci Rep 12(1):17474. https://doi.org/10.1038/s41598-022-21493-w
Martin R (2021) Rebuilding the economy from the Covid crisis: time to rethink regional studies? Reg Stud Reg Sci 8(1):143–161
Morin E (2006) Flat earth. Nueva Visión, Argentina
Nava A et al (2017) The impact of global environmental changes on infectious disease emergence with a focus on risks for Brazil. ILAR J 58(3):393–400
Patz JA et al (2004) Unhealthy landscapes: policy recommendations on land use change and infectious disease emergence. Environ Health Perspect 112(10):1092–1098
Pfenning-Butterworth A, Buckley LB, Drake JM, Farner JE, Farrell MJ, Gehman ALM, Davies TJ (2024) Interconnecting global threats: climate change, biodiversity loss, and infectious diseases. Lancet Planet Health 8(4):e270–e283
Rehbein B (2020) Capitalism and inequality. Sociedade e Estado 35:695–722
Rockström J, Steffen W, Noone K, Persson Å, Chapin III FS, Lambin E, Lenton T, Scheffer M, Folke C, Schellnhuber H, Nykvist B, de Wit AC, Hughes T, Leeuw S, Rodhe H, Sörlin S, Snyder P, Costanza R, Svedin U, Falkenmark M, Karlberg L, Corell R, Fabry V, Hansen J, Walker B, Liverman D, Richardson K, Crutzen P, Foley J (2009) Planetary boundaries: exploring the safe operating space for humanity. Ecol Soc 14(2):1–34
Royo-Bordonada MÁ (2019) Corporate capture of public health. Revista De Bioética y Derecho 45:25–41
Ruuska T (2017) Capitalism and the absolute contradiction in the Anthropocene. In: Sustainability and peaceful coexistence for the Anthropocene, pp 51–67. Routledge
Schmelzer M, Vetter A, Vansintjan A (2022) The future is degrowth—a guide to a world beyond capitalism, Ed. Verso, London, UK
Schor JB (2005) Prices and quantities: unsustainable consumption and the global economy. Ecol Econ 55(3):309–320. https://doi.org/10.1016/j.ecolecon.2005.07.030
Sell SK, Williams OD (2020) Health under capitalism: a global political economy of structural pathogenesis. Rev Int Polit Econ 27(1):1–25. https://doi.org/10.1080/09692290.2019.1659842
Sigler T, Mahmuda S, Kimpton A, Loginova J, Wohland P, Charles-Edwards E, Corcoran J (2021) The socio-spatial determinants of COVID-19 diffusion: the impact of globalization, settlement characteristics and population. Glob Health 17(1):56. https://doi.org/10.1186/s12992-021-00707-2
Soeder DJ (2021) Fossil fuels and climate change. In: Fracking and the environment. Springer, Cham. https://doi.org/10.1007/978-3-030-59121-2_9
Spangenberg JH (2017) Hot air or comprehensive progress? a critical assessment of the SDGs. Sustain Dev 25(4):311–321
Spash C (2017) Routledge handbook of ecological economics—nature and society. New York, USA, Routledge
Spash C (2021) Apologists for growth: passive revolutionaries in a passive revolution. Globalizations 18(7):1123–1148. https://doi.org/10.1080/14747731.2020.1824864
Spash, C. and Guisan, A. (2021). A future social-ecological economics. Real World Econ Rev 96(July):220–223. http://www.paecon.net/PAEReview/issue96/SpashGuisan96.pdf
Spash CL, Méral P, Petit O (2023) A journey into the foundations and transformative implications of social ecological economics: an interview with Clive Spash. Socio-économie écologique et dynamiques territoriales 14(1). https://doi.org/10.4000/developpementdurable.22652
Stefania-Barca (2018) In defense of degrowth: opinions and minifestos/doughnut economics: seven ways to think like a 21st-century economist. Local Environ 23(3):378–381. https://doi.org/10.1080/13549839.2017.1399997
Sullivan D, Hickel J (2023) Capitalism and extreme poverty: a global analysis of real wages, human height, and mortality since the long 16th century. World Dev 161:106026
The Lancet Planetary Health (2024) Beyond incrementalism. Lancet Planet Health 8(4):e213. https://doi.org/10.1016/S2542-5196(24)00043-3

Warren K (2000) Ecofeminist philosophy: a western perspective on what it is and why it matters. Rowman & Littlefield

Williams R (2016) The privileges of wealth: rising inequality and the growing racial divide. Routledge.

Windegger F, Spash CL (2022) Reconceptualising Freedom in the 21st Century: neoliberalism vs. degrowth. New Polit Econ 1–20. https://doi.org/10.1080/13563467.2022.2149719.

Zipperer WC, Foresman TW, Walker SP, Daniel CT (2012) Ecological consequences of fragmentation and deforestation in an urban landscape: a case study. Urban Ecosyst 15:533–544

Social Entrepreneurship in Tourism: A Systemic Perspective

5

Ricardo Tejeida-Padilla, Zeltzin Pérez-Matamoros, Roberto Patiño-Abuela, and Isaías Badillo-Piña

Abstract

Social entrepreneurship in tourism emerges as a transformative alternative that promotes more equitable economic development while safeguarding cultural identity and fostering balance with the environment. By integrating principles of social and environmental responsibility into tourism activities, these initiatives empower local communities and encourage social and economic inclusion from a participatory perspective. This research aims to provide a synthesis of the relationship between social entrepreneurship and tourism dynamics from a systemic approach, through a systematic and systemic literature review. To this end, the Total Systems Intervention (TSI) is incorporated, enabling the complexity of tourism systems to be addressed in their social, economic, and environmental dimensions. The findings demonstrate that a systemic perspective allows for a comprehensive understanding of this emerging field and highlights its potential to generate a positive and lasting social impact.

R. Tejeida-Padilla (✉) · R. Patiño-Abuela
SEPI EST, National Polytechnic Institute, Mexico City, Mexico
e-mail: rtejeidap@ipn.mx

Z. Pérez-Matamoros · I. Badillo-Piña
SEPI ESIME-Zacatenco, National Polytechnic Institute, Mexico City, Mexico

Keywords

Tourism · Innovation · Sustainability · Social entrepreneurship · Complex systems

5.1 Introduction

Social entrepreneurship stands out as one of the key factors for driving regional economies and promoting positive social change within society. This assertion is supported by the innovative integration of resources applied in programs, initiatives, organizations, products, or services created through social enterprises (Dahles et al. 2020; Fauzi et al. 2022; Gupta et al. 2020). The phenomenological nature of social entrepreneurship highlights its capacity to generate significant impact across various domains, and tourism is no exception to this phenomenon.

When tourism is viewed from a systemic perspective, its complex nature becomes evident, encompassing a wide range of interactions among different elements such as natural, cultural, economic, and social resources (Tejeida et al. 2016). This complexity is particularly apparent in regions that receive a considerable number of tourists, where socio-environmental dynamics are in constant flux.

The continuous growth of tourism presents a series of opportunities and challenges for host communities, who must sustainably manage and leverage local resources while meeting the

P. Pérez-Hernandez and R. Huerta Cuervo (eds.), *Social Well-Being and the Challenges in the Sustainable Development Goals*, Sustainable Development Goals Series,
https://doi.org/10.1007/978-3-032-14294-8_4

evolving demands of visitors (Anthonisz and Mason 2019; Zhang and Zhang 2023).

The interaction between social entrepreneurship and tourism represents a holistic approach to sustainable development, transcending economic growth and focusing on comprehensively addressing the social and environmental needs of local communities (de Lange and Dodds 2017; Li et al. 2022). This synergy between social entrepreneurship and tourism dynamics enables the creation of initiatives that not only generate financial benefits but also promote social well-being, equity, and the preservation of the natural and cultural environment (Singh and Reji 2020; Utami et al. 2023).

Community involvement in tourism dynamics encompasses four main dimensions: economic, social (internal relations and collective participation), informational, and educational. Applying this holistic approach also recognizes other dimensions, such as the empowerment of community members, which significantly contributes to strengthening their sense of individual autonomy (Tong et al. 2024). Rappaport (1984) suggests that empowerment has several levels of analysis; at the individual level, empowerment includes participatory behavior, motivation to exercise control, and feelings of efficacy and control.

Active community participation in tourism planning and management enables local residents to make informed decisions and take a leading role in the development of their environment, strengthening their autonomy and leadership capacity. By integrating principles of social and environmental responsibility into tourism dynamics, projects can be promoted that contribute to the empowerment of local communities, fostering social and economic inclusion (Herutomo et al. 2022; Nugraheni et al. 2019). This generates a commitment to environmental conservation, encouraging sustainable practices for the protection of natural resources and the reduction of the negative impact generated by tourism activities when not properly managed.

The aim of this research is to develop a synthesis between social entrepreneurship and tourism from a systemic perspective, in order to identify how these initiatives contribute to sustainability, innovation, and inclusion in tourist destinations. To this end, a systematic and systemic literature review is conducted, integrating elements of the Total Systems Intervention (TSI) to understand the complexity of interactions among actors, resources, and practices in this emerging field.

5.2 Methodology

For the systematic literature review, explicit and systematic methods are used to collect and synthesize findings from research addressing the relevant systems of social entrepreneurship and tourism. The methodology is based on the taxonomy provided by the state-of-the-art matrix in conjunction with the PRISMA 2020 Statement matrix (Page et al. 2021), which enables effective management in grouping and classifying the reviewed literature, both by keywords and by the topics addressed.

In this case, the literature search was conducted considering research published from 2017 to January 2024, using the following group of keywords: tourism, social entrepreneurship, complex system, and participant relationships; the latter derived from the systemic perspective generated by TSI (Jackson 2003), as shown in Table 5.1.

The set of keywords was determined based on the conceptualization of tourism as a dynamic entity and social entrepreneurship, which explain and describe the diversity and complexity found in tourism, where socially oriented ventures are created. The search was conducted in two of the

Table 5.1 Eligibility criteria

Keywords	Selection process	Information sources	
		Databases	Indexing
Tourism, social entrepreneurship, innovation, sustainability, complex system, participant relationships	That the keywords are present in the titles and/or abstracts	Web of science and science direct	JCR and scopus

Source Own elaboration

most popular and widely used repositories by the academic community: ScienceDirect and Web of Science.

The filtering process was carried out in two stages. In Web of Science, a search was performed using groups of keywords with the Boolean operator "AND" since the database does not allow for the complete entry of all keywords in a single search. Meanwhile, in ScienceDirect, the search was conducted with all keywords together, also using the Boolean operator "AND" between each of them. Additionally, for the concepts of complex systems and participant relationships, derived from Systems Theory, the International Encyclopedia of Systems and Cybernetics (François 2004) was used, with the aim of synthesizing these concepts within the context of the research.

5.3 Results

The results obtained reveal an extensive body of literature on social entrepreneurship in the field of tourism and its relationship with innovation and sustainability, with a total of 52,993 articles identified in the Web of Science and ScienceDirect databases (Fig. 5.1), covering research published from 2017 to January 2024.

In the Web of Science database, a total of 52,275 articles were identified. Within this set, 1,356 articles contained the keywords: "tourism" and "social entrepreneurship"; 50,762 articles with the keywords: "sustainability" and "innovation"; 1,395 articles with the keywords: "complex system" and "relationship with participants"; 116 articles with the keywords: "tourism," "social entrepreneurship," "sustainability," and "innovation." Notably, only 2 articles addressed the intersection of "tourism," "social entrepreneurship," "sustainability," "innovation," and "complex system" as keywords; however, no articles were found that addressed all six keywords together.

On the other hand, in the ScienceDirect database, a total of 718 articles were found related to the keywords "tourism," "social entrepreneurship," "sustainability," "innovation," "complex system," and "participant relationship." These results confirm the growing interest in understanding the complex relationships among social

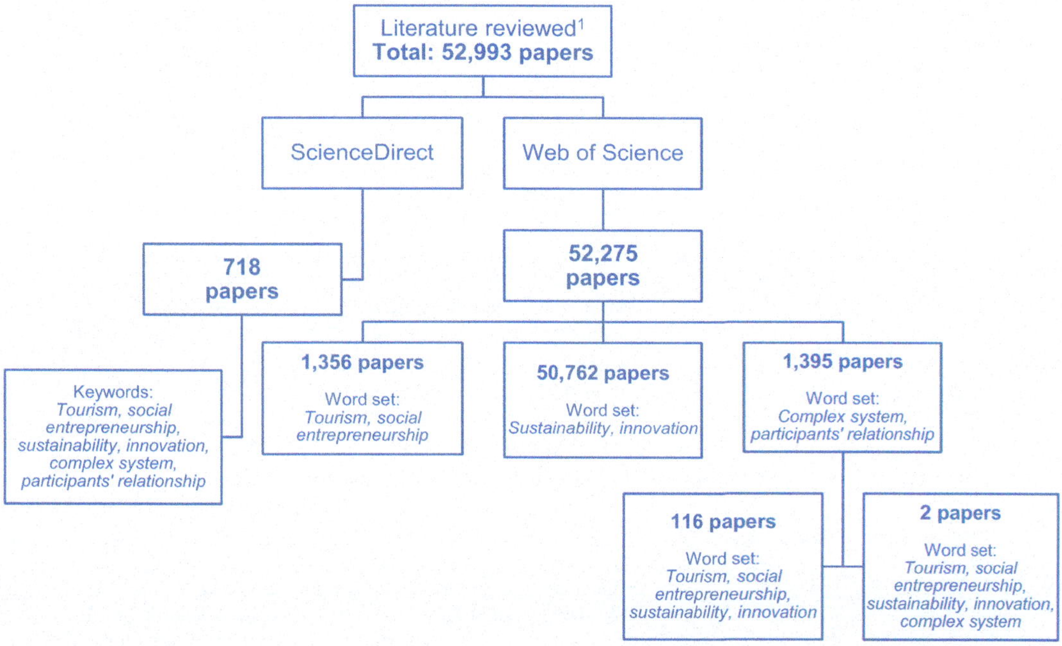

Fig. 5.1 Selection of articles based on keywords. [1]The subsets shown are not mutually exclusive. Instead, they represent progressive refinements of the initial total, based on specific keyword combinations. *Source* Own elaboration

entrepreneurship, tourism, sustainability, innovation, and the dynamics of stakeholder participation.

The literature review revealed that the concept of entrepreneurship has evolved over time. Exploring the roots and meaning of entrepreneurship is essential to understanding its scope and development throughout history. The term *entrepreneurship* originates from the French verb *entreprendre* and the German word *unternehmen*, both denoting the act of undertaking (Ansari and Mohd 2009).

The roots of the concept of entrepreneurship date back to classical economic thought in the eighteenth century, especially in France with the physiocrats, among whom Richard Cantillon stands out. Cantillon was the first to introduce the modern concept of entrepreneurship, describing it as the willingness of individuals to create new forms of intermediation, assuming the economic risk inherent in a new venture (Rodríguez 2009).

Since then, the conceptualization of entrepreneurship has undergone significant evolution. Table 5.2 presents a description of the various concepts developed by leading authors worldwide, highlighting the different contributions that have enriched the understanding of this phenomenon. It is also important to note that the multifaceted nature of entrepreneurship continues to challenge a universally accepted definition, underscoring the complexity and dynamism inherent in this field.

The evolution of the concept of entrepreneurship has shifted from definitions focused on

Table 5.2 Evolution of the concept of entrepreneurship

Year	Author	Term	Concept	Contribution
1780	Adam Smith	Projector/undertaker	An individual who assumes three roles: owner, manager, and risk-taker	It recognizes for the first time the connection between innovation and professional activity within entrepreneurship
1804	Jean-Baptiste Say	Entrepreneur	A visionary leader who takes risks and evaluates projects, mobilizing resources from low-performing areas to those with higher productivity	Emphasizes the importance of the entrepreneur's success for society as a whole, defining it as a "superior form of work"
1848	John Stuart Mill	Undertake	Entrepreneurship comprises three components of profit: interest, insurance, and supervisory wages. The central figure (*undertaker*) is someone who organizes and assumes the risks of entrepreneurship in exchange for these profits	Emphasized the significance of entrepreneurship in fostering economic growth
1890	Alfred Marshall	Entrepreneur	A natural leader, ready to act in uncertain conditions and endowed with a wide range of distinctive skills	Referenced leadership skills and added the factor of industrial organization
1911	Joseph A. Schumpeter	Entrepreneurship	Drives economic development through an innovative approach, and its relevance in the modern economy is widely acknowledged	Was the first to propose the concept of dynamic disequilibrium resulting from innovative entrepreneurship
1990	William Baumol	Entrepreneurship	A mechanism that enables economic growth	Theory of productive and unproductive entrepreneurship
1991	William D. Bygrave	Entrepreneurship	A process that involves establishing an organization in which activities, functions, and actions are undertaken to seize opportunities	Defined the characteristics and functions of the entrepreneur and outlined the entrepreneurial process through strategic planning

Source Own elaboration based on information from Baumol (1968), Bygrave and Hofer (1992), Elkjaer (1991), Fernández-Salinero and de la Riva (2014), McDaniel (2005), Saposnik et al. (1982)

ownership, management, and risk-taking to a broader perspective that views it as a comprehensive process of creating organizations aimed at seizing opportunities. Likewise, the entrepreneur is increasingly seen as a visionary leader, capable of mobilizing resources toward areas of greater productivity and fostering economic and social progress. Their impact on economic dynamics has become established as a key mechanism for strengthening regional economies, particularly through their ability to drive organizational innovation.

In contrast to this logic, which is primarily centered on the economic sphere, social entrepreneurship has emerged as a significant response to today's most complex social, economic, and environmental challenges (Akay 2022; Vig 2023), as it adopts a holistic approach that considers economic viability, community well-being, and environmental preservation (Kummitha et al. 2021). Among the distinctive features of social entrepreneurship in tourism are the ability to generate innovative, community-focused solutions, social inclusion, and a commitment to sustainability (Kumar 2023).

Social entrepreneurship also has great potential to generate positive impacts at various levels. Economically, it can drive local development by creating employment opportunities for community members. At the same time, these initiatives strengthen the social fabric and help preserve cultural identity by involving communities in decision-making and the collective management of tourism resources (Sochacka and Rzeszotarska-pałka 2021).

From a sustainability perspective, social entrepreneurship in tourism promotes responsible and environmentally respectful practices. These include biodiversity conservation, sustainable management of natural resources, and reducing the ecological footprint of tourism products and services. In turn, such initiatives can foster the diversification of tourism offerings through authentic products and services that value local culture and promote responsible tourism (Graça et al. 2023).

Figure 5.2 summarizes the main dynamics that emerge in tourism when driven by social entrepreneurship initiatives. These dynamics do not operate in isolation; rather, they are interwoven

Fig. 5.2 Interrelated elements shaping tourism through social entrepreneurship in complex environments. *Source* Author's own elaboration

within a complex environment where the elements serve as drivers of transformation.

Each of these elements contributes to a more inclusive, resilient, and socially engaged tourism experience. Furthermore, social entrepreneurship is shown to foster social and economic inclusion in tourist destinations by providing employment and training opportunities to marginalized or disadvantaged groups, such as women, youth, or people with disabilities. These initiatives empower local communities and enrich the tourism offering by diversifying the experiences available to visitors.

Thus, social entrepreneurship not only introduces new practices within tourism but also invites a rethinking of the underlying structures: who makes decisions, who benefits, and how resources are distributed. From this perspective, it becomes a tool for transforming traditional tourism (Figueroa-Domecq et al. 2023; Li et al. 2022).

In summary, the findings show that social entrepreneurship is not only an emerging trend but also a concrete strategy for transforming tourism activity from a sustainable perspective. By prioritizing social impact over economic profit, these initiatives adopt purpose-driven business models capable of generating real and lasting value (Podmetina et al. 2022; Skivko 2021), offering innovative solutions to current and future challenges.

5.4 Conclusions

Understanding and addressing these aspects from a systemic perspective is essential for advancing toward more balanced tourism development that is connected to the realities of local territories. To date, no studies have been found that characterize social entrepreneurship within the dynamics of tourism as a complex system, nor that consider the relationships among participants in social entrepreneurship.

The convergence of these topics highlights the need to adopt a holistic approach to understand economic and social development within their inherent complexity. Recognizing and integrating a systemic perspective allows for a more comprehensive approach to contemporary challenges, connecting dimensions that are often addressed separately and generating more coherent and sustainable solutions.

More than an economic strategy, social entrepreneurship in tourism has become a key driver for generating collective well-being and protecting local territories. Its true value lies in its ability to maintain homeostasis among development, community rootedness, and respect for the environment. In this context, sustainability is not just a principle, but a complementary pillar that guides these initiatives toward minimizing environmental impact and generating social and economic benefits that endure over time.

Innovation also plays a crucial role in the creation and development of social enterprises in the tourism sector, facilitating the generation of creative and sustainable solutions to existing challenges. Understanding the dynamics of stakeholder participation in these processes is fundamental to ensuring the effectiveness and positive impact of social entrepreneurship initiatives in tourism.

Social enterprises that focus on addressing specific challenges within the tourism sector—such as waste management, cultural heritage conservation, and the promotion of responsible tourism—are clear examples of the potential of entrepreneurship to solve current problems in tourism. Likewise, a transdisciplinary and collaborative approach to addressing this complexity fosters improvements in reality at multiple levels simultaneously.

Social entrepreneurship is expected to continue playing an important role in the positive transformation of tourism dynamics, thereby contributing to the achievement of the Sustainable Development Goals and the construction of a fairer and more equitable tourism sector for all.

Acknowledgements This research is supported by the Instituto Politécnico Nacional through transdisciplinary research project SIP3420, modules 20254027 and 20254253, by the Secretaría de Investigación y Posgrado and the Secretaría de Ciencia, Humanidades, Tecnología e Innovación.

References

Akay B (2022) The relationship between tourism development, satisfaction, support to tourism and tourism entrepreneurship in emerging tourism destination: residents' perspective support to tourism and tourism entrepreneurship in emerging keywords. https://doi.org/10.18488/31.v9i1.3078

Ansari VA, Mohd AU (2009) Role of venture capital in spurring innovation and entrepreneurship V(1):114–119

Anthonisz A, Mason G (2019) Reinventing tourism: the Dubai phenomenon 11(3):279–286. https://doi.org/10.1108/WHATT-01-2019-0004

Baumol WJ (1968) Entrepreneurship in economic theory. Am Econ Rev 58(2):64–71

Bygrave WD, Hofer CW (1992) Theorizing about Entrepreneurship. Entrep Theory Pract 16(2):13–22. https://doi.org/10.1177/104225879201600203

Dahles H, Khieng S, Verver M, Manders I (2020) Social entrepreneurship and tourism in Cambodia: advancing community engagement. J Sustain Tour 28(6):816–833. https://doi.org/10.1080/09669582.2019.1706544

de Lange D, Dodds R (2017) Increasing sustainable tourism through social entrepreneurship. Int J Contemp Hosp Manag 29(7):1977–2002. https://doi.org/10.1108/IJCHM-02-2016-0096

Elkjaer JR (1991) The entrepreneur in economic theory. An example of the development and influence of a concept. Hist Eur Ideas 13(6), 805–815. https://doi.org/10.1016/0191-6599(91)90143-M

Fauzi MA, Muhamad Tamyez PF, Kumar S (2022) Social entrepreneurship and social innovation in ASEAN: past, present, and future trends. J Soc Entrep 1–23. https://doi.org/10.1080/19420676.2022.2143870

Fernández-Salinero C, de la Riva B (2014) Entrepreneurial mentality and culture of entrepreneurship. Procedia Soc Behav Sci 139:137–143. https://doi.org/10.1016/j.sbspro.2014.08.044

Figueroa-Domecq C, Kimbu A, de Jong A, Williams AM (2023) Sustainability through the tourism entrepreneurship journey: a gender perspective. In: Gender and tourism sustainability. Routledge, pp 107–130.

François C (2004) International encyclopedia of systems and cybernetics. De Gruyter Saur

Graça A, Morais I, Silva H, Guerra C, Manso J, Marcos L, Carlos C (2023) Douro wine-tourism engaging consumers in nature conservation stewardship: an immersive biodiversity experience, p 03003

Gupta P, Chauhan S, Paul J, Jaiswal MP (2020) Social entrepreneurship research: a review and future research agenda. J Bus Res 113:209–229. https://doi.org/10.1016/j.jbusres.2020.03.032

Herutomo A, Dhewanto W, Prasetio EA (2022) Social entrepreneurship and sustainable development goals: a conceptual framework. Eur Conf Innov Entrepren 17(1):275–283. https://doi.org/10.34190/ecie.17.1.505

Jackson M (2003) Systems thinking—creative holism for managers. Kybernetes 33(8). https://doi.org/10.1108/k.2004.06733hae.001

Kumar D (2023) Social innovation for sustainability and social inclusion. https://doi.org/10.1177/2319510X221144657

Kummitha HR, Kolloju N, Jancsik A, Szalók ZC (2021) Can tourism social entrepreneurship organizations contribute to the development of ecotourism and local communities: understanding the perception of local communities. Sustainability 13(19):11031. https://doi.org/10.3390/su131911031

Li X, Abbas J, Dongling W, Baig NUA, Zhang R (2022) From cultural tourism to social entrepreneurship: role of social value creation for environmental sustainability. Front Psychol 13(July). https://doi.org/10.3389/fpsyg.2022.925768

McDaniel BA (2005) A contemporary view of Joseph A. Schumpeter's theory of the entrepreneur. J Econ Issues 39(2):485–489. https://doi.org/10.1080/00213624.2005.11506826

Nugraheni AIP, Priyambodo TK, Sutikno B, Kusworo HA (2019) Defining social sustainability and socially conscious tourist in sustainable tourism development. J Bus Hosp Tourism 5(2):264. https://doi.org/10.22334/jbhost.v5i2.172

Page MJ, McKenzie JE, Bossuyt PM, Boutron I, Hoffmann TC, Mulrow CD, Shamseer L, Tetzlaff JM, Akl EA, Brennan SE, Chou R, Glanville J, Grimshaw JM, Hróbjartsson A, Lalu MM, Li T, Loder EW, Mayo-Wilson E, McDonald S, Moher D (2021) The PRISMA 2020 statement: an updated guideline for reporting systematic reviews. Int J Surg 88(March):105906. https://doi.org/10.1016/j.ijsu.2021.105906

Podmetina D, Albats E, Rosei F, Kautto D (2022) Innovation and entrepreneurship for social goals and sustainability in developing countries. In: Proceedings of the annual Hawaii international conference on system sciences, pp 5295–5303. https://doi.org/10.24251/HICSS.2022.645

Rappaport J (1984) Studies of empowerment: introduction to the issue, prevention in human issue

Rodríguez A (2009) New perspectives for understanding business entrepreneurship. Pensamiento & Gestión 26:94–119. http://www.scielo.org.co/pdf/pege/n26/n26a05.pdf

Saposnik R, Horwich G, Quirk JP (1982) Essays in contemporary economics. South Econ J 49(2):591. https://doi.org/10.2307/1058527

Singh A, Reji EM (2020) Social entrepreneurship and sustainable development. Taylor & Francis

Skivko M (2021) Digital technologies, social entrepreneurship and governance for sustainable development. Res Soc Change 13(1):165–173. https://doi.org/10.2478/rsc-2021-0016

Sochacka E, Rzeszotarska-pałka M (2021) Social perception and urbanscape identity of flagship cultural developments in Szczecin (in the Re-Urbanization Context)

Tejeida R, Coria EC, Juárez AB (2016) Systems thinking and tourism. Editorial Miguel Ángel Porrúa

Tong J, Li Y, Yang Y (2024) System construction, tourism empowerment, and community participation: the sustainable way of rural tourism development. Sustainability (Switzerland) 16(1). https://doi.org/10.3390/su16010422

Utami DD, Dhewanto W, Lestari YD (2023) Rural tourism entrepreneurship success factors for sustainable tourism village: evidence from Indonesia. Cogent Bus Manage 10(1). https://doi.org/10.1080/23311975.2023.2180845

Vig S (2023) Sustainable development through sustainable entrepreneurship and innovation: a single-case approach 19(7):1196–1217. https://doi.org/10.1108/SRJ-02-2022-0093

Zhang J, Zhang Y (2023) Tourism and regional carbon emissions: city-level evidence from China. 78(3):888–906. https://doi.org/10.1108/TR-08-2022-0389

Green Finance and Its Relationship with Economic Growth in Latin American Countries: A Critical Review

6

Pablo Emilio Escamilla-García
and José Antonio Morales-Castro

Abstract

This chapter provides an exploration of the growth of green finance in the Latin American region. Through a critical review of the literature and analysis of key data, it addresses the essential elements that allow for the identification of the current status of green finance in selected economies of the region. The discussion begins with a detailed analysis of green finance and its conceptualization in the Latin American context. Various financial strategies and instruments that have emerged to align investments with sustainable and responsible practices are examined. In addition, emerging trends in the region are highlighted, identifying specific opportunities and challenges faced by stakeholders in the financial sector. A fundamental part of the critical review focuses on assessing the current status of green finance in the five main economies of Latin America. To this end, the regulatory framework applicable in each country is analyzed, the main issuers of green bonds, as well as key projects that incentivize sustainable finance. This critical approach enables a more nuanced understanding of how green finance can serve as either a catalyst or a constraint for economic growth in the region. The importance of active participation by governments, financial institutions, and civil society in shaping an enabling environment for the development of green finance is emphasized. Ultimately, the chapter concludes by highlighting the opportunities that green finance offers to reshape the economic landscape of Latin America in a sustainable manner. The importance of addressing the identified challenges is underscored, and future perspectives are outlined, emphasizing the need for comprehensive collaboration to fully leverage the transformative potential of green finance in the region.

Keywords

Green finance · Green bonds · Latin America · Sustainability

6.1 Introduction

In recent decades, the concept of sustainable finance has emerged as a fundamental pillar in global economic and financial management, gaining particular relevance in Latin America. The region has undergone a significant trans-

P. E. Escamilla-García (✉)
Instituto Politécnico Nacional - Unidad Profesional Interdisciplinaria de Ingeniería y Ciencias Sociales y Administrativas, Mexico City, Mexico
e-mail: peescamilla@ipn.mx

J. A. Morales-Castro
Institito Politécnico Nacional - Escuela Superior de Comercio y Administración Unidad Tepepan, Mexico City, Mexico
e-mail: jmorales@ipn.mx

P. Pérez-Hernandez and R. Huerta Cuervo (eds.), *Social Well-Being and the Challenges in the Sustainable Development Goals*, Sustainable Development Goals Series,
https://doi.org/10.1007/978-3-032-14294-8_5

formation in the perception and implementation of financial practices that integrate environmental, social, and corporate governance considerations (Correa 2007). This shift toward sustainable finance not only responds to a growing awareness of the importance of addressing environmental and social challenges, but also reflects a strategic response to the increasing demand for responsible investment from investors and society at large. The purpose of this article is to analyze the evolution of the concept of sustainable finance in Latin American countries and to explore how this transformation has positively impacted the achievement of the United Nations Sustainable Development Goals (SDGs). As the region faces persistent challenges such as poverty, inequality, environmental degradation, and climate change, the role of sustainable finance has become essential for fostering inclusive and sustainable economic development (Samaniego and Schneider 2023). This study examines the trends, challenges, and opportunities in the adoption of sustainable financial practices in Latin America, highlighting specific cases of countries that have led the way in this transition. It also explores how financial institutions, companies, and governments have collaborated to integrate sustainability principles into their strategies and operations. By linking these advances to the achievement of the SDGs, the aim is to shed light on the tangible contribution of sustainable finance to improving key indicators related to poverty eradication, gender equality, climate action, and other fundamental objectives for sustainable development. As this analysis progresses, the intention is to identify common patterns, lessons learned, and potential areas for improvement in the landscape of sustainable finance in Latin America. In doing so, the study seeks to provide a solid foundation for future research, as well as to inform policymakers, investors, and key stakeholders about the strategic importance of integrating sustainability into the region's financial fabric.

In the twenty-first century, the commitment to sustainability and the mitigation of environmental impact have acquired unprecedented importance on a global scale. Within this context, green finance has emerged as a key catalyst for driving the transition toward a more sustainable economy. Latin America, a region rich in natural resources and diversity, has not been immune to this evolution (Herrera 2023). The integration of financial practices that promote environmental sustainability has become a central topic in the economic and political discourse of various countries in the region. The relationship between green finance and economic growth in Latin America represents a strategic and highly relevant field of study today. As the region faces persistent challenges, from natural resource management to climate change adaptation, understanding how sustainable finance impacts economic development becomes essential. This study is justified by the need to provide a comprehensive view of the dynamics between green finance and economic growth, enabling key stakeholders in the region to make informed decisions to advance a sustainable development agenda. This review aims to provide a comprehensive framework for understanding the crucial role that green finance plays in the economic landscape of Latin America, thereby contributing to informed decision-making and the design of strategic policies for long-term sustainable development.

6.2 Theoretical Framework

6.2.1 Background of Green Finance

The concept of green finance emerged in response to growing concerns about global environmental and climate issues. Its aim is to promote sustainability and mitigate negative environmental impacts through responsible financial practices. The origins of green finance can be traced in the development of corporate social responsibility (CSR) and environmental awareness.

- Development of Corporate Social Responsibility (CSR): In the last decades of the twentieth century, companies began to recognize the importance of considering social and environmental factors in their operations. CSR initially focused on ethical and social aspects, but over time expanded to address environmental issues as well (Chancay-Cedeño 2016).

- Kyoto Protocol (1997): This international treaty marked a turning point in global awareness of climate change. It established commitments to reduce greenhouse gas emissions. Following this protocol, financial institutions began to consider environmental impact in their investment decisions (Dessai et al. 2003).
- Brundtland Report (1987): The United Nations World Commission on Environment and Development published this report, which introduced the concept of sustainable development. The report emphasized the importance of meeting the needs of the present generation without compromising the ability of future generations to meet their own needs (Keeble 2007).
- Equator Principles (2003): The financial sector adopted the Equator Principles, which set standards for assessing and managing environmental and social risks in financed projects. This marked a significant step toward integrating environmental criteria into financial decision-making (Gallardo 2006).
- UN Climate Summits: United Nations climate change conferences, such as COP21 in Paris in 2015, have driven initiatives to mobilize funds toward sustainable investments. The Paris Agreement was a major milestone in the fight against climate change (Elbers 2013).
- Development of Green Financial Instruments: As environmental awareness grew, specific financial products emerged, such as green bonds and sustainable funds, designed to finance ecological and sustainable projects.

Today, green finance is gaining momentum with the growing demand for sustainable investments. Investors, companies, and governments recognize the importance of aligning financial activities with environmental objectives to address environmental challenges and promote sustainable development.

6.2.2 Definition of Green Finance

The study of green finance has become established as a multidisciplinary field that merges economic principles with environmental sustainability. At its core, it refers to financial and investment practices aimed at generating a positive impact on the environment and society (Gafoor et al. 2024). Since its conceptualization, the term has evolved beyond a mere focus on mitigating environmental risks to embrace the idea of financial opportunities aligned with ethical and sustainable criteria. In this context, green finance encompasses a wide range of financial instruments and investment strategies designed to promote resource efficiency, reduce carbon emissions, and encourage socially responsible business practices (Wu et al. 2024).

6.2.3 Theoretical Foundation of Green Finance

Green finance is grounded in several economic and financial theories and concepts that seek to integrate environmental and social considerations into financial decision-making. Some of the key theories underpinning the existence of green finance and their relationship to the concept are summarized in Table 6.1.

6.2.4 Main Principles and Instruments of Green Finance

- Socially Responsible Investments (SRI): Green finance is grounded in the concept of SRI, where investors seek financial returns while also considering environmental, social, and governance (ESG) factors (Viñuales and Pueyo 2002). The integration of ESG criteria into investment decision-making is a fundamental principle.
- Green and Sustainable Bonds: Green bonds, as well as sustainable bonds, are key financial instruments. These instruments channel

Table 6.1 Relationship between green finance and economic-financial theories

Theory	Concept	Relationship
Social capital theory	Social capital refers to social networks, norms, trust, and cooperation that facilitate collective action for the common good	Green finance seeks investments that not only generate financial returns but also contribute to social and environmental well-being, aligning with the principles of sustainability and social responsibility
Corporate social responsibility (CSR) theory	CSR implies that companies should operate ethically and contribute to sustainable development, considering not only their economic benefits but also the social and environmental impacts of their activities	In the context of green finance, CSR translates into investments that support companies and projects with sustainable practices, helping to drive the green economy
Common good economy theory	This theory proposes measuring economic success based on human and environmental well-being, rather than solely through traditional financial indicators	Green finance shares the goal of contributing to overall well-being, as it seeks investments that generate benefits for both society and the environment
Socially responsible investment (SRI) theory	SRI involves selecting investments not only based on financial criteria, but also on ethical, social, and environmental considerations	Green finance adopts SRI principles by focusing on investments that have a positive impact on the environment and society, promoting sustainability
Sustainable development theory	Proposed by the World Commission on Environment and Development in 1987, sustainable development seeks to meet present needs without compromising the ability of future generations to meet their own needs	Green finance aligns with sustainable development by directing investments toward projects that balance economic, social, and environmental needs
Shared value theory	Developed by Michael Porter and Mark Kramer, this theory holds that companies can achieve economic success while also contributing to social and environmental development	Green finance seeks to generate shared value by financing projects that benefit both society and investors, creating a positive impact across multiple dimensions

Source Own elaboration

funds toward projects with measurable environmental and social benefits, such as renewable energy, energy efficiency, and sustainable infrastructure projects (Garayoa and García 2018).

- Impact Investments: The impact investment approach seeks to generate social and environmental benefits in addition to financial returns. This approach goes beyond mere risk mitigation, aiming to actively contribute to the achievement of sustainable goals (Contreras, Pedraza and Martínez 2017).
- Sustainable Investment Funds: The creation of investment funds specifically dedicated to sustainable assets and companies committed to responsible practices is another essential component of Green Finance (De Santiago 2020).

6.3 Global Context of Green Finance

The global landscape of green finance has experienced exponential growth in recent years. International organizations, financial institutions, and governments have played a key role in promoting standards and regulations to drive the adoption of sustainable financial practices (Gafoor et al. 2024). International agreements, such as the United Nations Principles for Responsible Banking and the Sustainable Development Goals (SDGs), have provided guiding frameworks that have influenced the direction of green finance worldwide. Growing awareness of climate risks and the need for a transition to a low-carbon economy have led to a significant increase in the mobilization of financial resources toward projects and companies aligned with environmental and social objectives (Zhang et al. 2023).

Specifically, the global context of green finance is unfolding in a scenario where increasing environmental awareness and concern about climate change have driven the search for sustainable solutions across various sectors of society. In this context, green finance emerges as a strategic approach that integrates environmental considerations into financial decision-making, addressing the critical need to mitigate the negative impacts of economic activities on the environment (Wu et al. 2024).

Public and governmental pressure to address environmental challenges has led to a paradigm shift in the financial world. Financial institutions, investors, and companies are increasingly recognizing the importance of incorporating environmental, social, and governance (ESG) criteria into their strategies. This shift is supported by growing evidence that sustainable practices are not only ethical but also profitable in the long term (Liu et al. 2023; Xu et al. 2024).

Within the framework of green finance, financial instruments such as green bonds, sustainable loans, and other products that channel resources toward projects with positive environmental impact are particularly noteworthy. In addition, the adoption of environmental standards and certifications provides a unified framework for evaluating and measuring the sustainable performance of investments (Yi et al. 2023). This approach seeks not only to generate financial returns, but also to contribute to environmental conservation and promote the transition to a cleaner and more sustainable economy.

Globally, various governmental initiatives and international agreements support the advancement of green finance. The Agenda 2030 United Nations Agenda and the Paris Agreements are key examples urging the international community to work together to address environmental issues through sustainable financing. In this context, green finance is not only seen as a tool to mitigate risks associated with climate change, but also as an opportunity to drive innovation and competitiveness in the global economy (La Torre et al. 2024).

This global context provides the backdrop for understanding the intersection between green finance and economic growth in Latin American countries, highlighting the importance of assessing how these trends translate into the specific context of different Latin American nations.

6.4 Evolution of Green Finance in Latin America

Latin America, a region rich in natural resources and biological diversity, faces a crucial crossroads at the intersection of economic development and environmental sustainability. In this context, green finance has emerged as an essential catalyst for addressing environmental challenges while promoting economic growth (McCauley et al. 2023). This geographic region faces increasing pressure to balance economic development with environmental conservation. The exploitation of natural resources has historically driven economic growth, but it has also contributed to environmental degradation (Castro 2007). In this context, green finance offers a strategic response to channel investments into projects that foster sustainability and mitigate negative impacts.

Several countries in the region have made progress in promoting green finance through the implementation of policies and regulations. Initiatives such as the issuance of green bonds and the creation of sustainable funds reflect governmental commitment to advancing the transition toward a greener economy (Ospina et al. 2023). However, disparities in development levels and implementation capacity still pose challenges to achieving uniform adoption across the region (Mendoza et al. 2021). Despite progress, Latin America faces significant barriers to the adoption of green finance. The lack of information and financial education, the perception of higher risks in sustainable investments, and the need for technical infrastructure to assess sustainable projects are challenges that must be addressed to encourage greater participation from the financial sector in this initiative (Stanley 2021). The adoption of green finance not only responds to environmental imperatives but also presents economic opportunities. Latin America can position itself as a leader in sustainable food production, renewable energy, and biodiversity conservation (Hwang 2023). Active participation in this field will not only strengthen environmental resilience but also enhance economic competitiveness on a global scale.

The landscape of green finance in Latin America reflects a strategic crossroads in the pursuit of a balance between economic growth and environmental preservation. Although significant progress has been made, challenges remain that require collective action from governments, businesses, and citizens. The adoption of green finance is not only a necessary response to the environmental crisis but also an opportunity to forge a sustainable and prosperous future for the region. International collaboration and ongoing commitment are essential to overcoming challenges and paving the way toward greener and more equitable development in Latin America.

6.5 Driving Factors and Barriers

Environmental concerns have led to the development and implementation of financial strategies aimed at mitigating these issues. Within the Latin American region, as previously discussed, green finance has represented a viable option for integrating economic and productive variables with environmental awareness and the common good. To achieve this, different Latin American economies have leveraged various driving factors to increase the scale and impact of green finance. Nevertheless, a range of barriers and obstacles have also emerged, hindering or limiting the growth of this sector. Below, some of the key driving factors and barriers for green finance in the region are discussed.

6.5.1 Driving Factors

1. Environmental Awareness and Social Commitment: The rise in environmental awareness among the Latin American population has been a key factor driving demand for sustainable financial practices. Social pressure to address environmental issues has led to a shift in perceptions of corporate and governmental responsibility, generating momentum for the adoption of financial practices aligned with sustainability (ECLAC 2023).
2. Government Commitments and Regulatory Framework: Several governments in Latin America have demonstrated explicit commitment to sustainability through the creation of regulatory frameworks that support green finance. The issuance of green bonds and the promotion of tax incentives for sustainable projects are examples of government measures that have stimulated growth in this sector (Guerrero 2021).
3. Economic Potential and Investment Opportunities: The region offers vast opportunities for sustainable investments in sectors such as renewable energy, sustainable agriculture, and biodiversity conservation. Investors and financial institutions see these opportunities as a way to achieve attractive financial returns while contributing to the mitigation of negative environmental impacts (Veltmeyer 2022).

6.5.2 Barriers

1. Lack of Knowledge and Financial Education: One of the most significant barriers in the region is the lack of knowledge and financial education regarding green finance. Both investors and financial institutions often lack the necessary understanding of the benefits and risks associated with sustainable investments, thus limiting their participation (McCauley et al. 2023).
2. Risk Perception: The perception of higher risks associated with sustainable investments compared to traditional ones has been an obstacle. The lack of a track record and robust data for sustainable projects can lead to investor reluctance to fully commit to these investments (Stanley 2021).
3. Need for Technical Infrastructure: Effective assessment of sustainable projects requires

specialized technical infrastructure. The lack of capacity to adequately evaluate the feasibility and environmental impact of these projects can be a significant barrier to the financial sector's participation (Hwang 2023).

The above shows that, in Latin America, both driving factors and barriers jointly influence the adoption of green finance. While environmental awareness and government commitments provide positive momentum, barriers such as lack of knowledge and risk perception must be addressed to enable broader and more effective development of green finance in the region. Collaboration among governments, financial institutions, and civil society will be essential to overcome these barriers and fully leverage the opportunities offered by green finance in Latin America.

6.6 Case Studies in Selected Latin American Countries

6.6.1 Argentina

6.6.1.1 Regulatory Framework

Regarding legislation in this area, according to information from the Ministry of Economy (2021), Argentina has various regulatory frameworks that have supported the growth of green finance. For example, the National Promotion Regime for the Use of Renewable Energy Sources for Electricity Production Act (Law 27.191) establishes a framework for the promotion and development of renewable energy sources. This has been key to the development of renewable energy projects in the country and could impact the financing of sustainable initiatives. The Capital Markets Law (Law 26.831) encourages the participation of the capital market in infrastructure and sustainable development projects. Financial instruments linked to renewable energy and sustainability projects may be influenced by this regulation. It should also be noted that Argentina has participated in international agreements related to sustainability and climate change, such as the Paris Agreement. The commitments made in these forums may influence the direction of sustainable financial policies and practices in the country. Additionally, some provinces and cities in Argentina have implemented specific regulations and commitments in the field of sustainability and the environment. These commitments can affect companies and projects at the local level.

6.6.1.2 Main Green Bond Issuers and ESG Stock Indices

In Argentina, there has been interest in sustainability and green finance projects, although the issuance of green bonds specifically was not as common as in some other markets (Trina 2019). Notable cases include Yacimientos Petrolíferos Fiscales (YPF), the state-owned oil and gas company, which has shown interest in projects related to renewable energy (Ferrer 2013). The company has been diversifying its operations toward more sustainable energy sources and may consider green finance initiatives in the future. Banco Galicia carried out the first green bond issuance to raise \$100 million to expand its environmental efficiency loan program (Municipality of Mendoza 2023). This was the first green bond issued by a private financial institution in Argentina. Additionally, Pampa Energía, an Argentine energy company, has been involved in electricity generation and may explore opportunities in renewable energy and sustainable projects; for example, in 2022, Pampa Energía issued its first green bond for \$3.107 billion to finance new investments in wind energy.

Stock exchanges play an important role in shaping sustainable markets through sustainable indices. Sustainable indices have become major catalysts for sustainable development, as they promote investment in companies that meet ESG requirements. In this regard, there is a range of sustainable indices and ESG information and rating agencies at both the international and Latin American levels (Inderst & Stewart, 2018). An index is a group or basket of instruments and/or financial securities that represent and measure the performance of a specific market, asset class, sector, or investment strategy. In Argentina, the ByMA Sustainability Index was created in 2018. This index was developed with the technical support and *know-how* of the Inter-American Development Bank and is based on the Index-Americas methodology. The index aims to identify and recognize market-leading companies for their sustainability practices. It evaluates issuers' performance across four pillars: environment, social, corporate governance, and development (contribution to the SDGs). According to the website of the Argentine Stock Exchange and Markets (BYMA), the index is composed of 20 companies (BVA 2023): Banco Hipotecario S.A., Banco Macro S.A., Banco BBVA Argentina S.A., Bolsas y Mercados Argentinos S.A., Central Puerto S.A., Enel Generación Costaneras S.A., Grupo Clarín S.A., Grupo Financiero Galicia S.A., Grupo Supervielle S.A., Holcim Argentina S.A., Irsa Inversiones y Representaciones S.A., Laboratorios Richmond S.A.C.I.F., Loma Negra Compañía Industrial Argentina S.A., Molinos Agro S.A., Molinos Río de la Plata S.A., Pampa Energía S.A., S.A. San Miguel A.G.I.C.I. y F., Telecom Argentina S.A., Transportadora de Gas del Sur S.A., and YPF S.A.

6.6.1.3 Projects for Green Finance

With regard to environmental projects that promote sustainable investments, the energy sector stands out. As of 2023, several major projects have been recorded, such as the Villalonga Wind Farm (Buenos Aires), Rawson Wind Farm (Chubut), Cauchari Solar Park (Jujuy), Corti Wind Farm (Buenos Aires), Miramar Wind Farm (Buenos Aires), and Cañada Honda Solar Park (San Juan) (Longo et al. 2022). These projects represent significant efforts in infrastructure development for a sustainable energy transition. Additionally, the livestock sector in Argentina has implemented sustainable projects, such as rotational grazing, where animals are moved between different pasture plots, allowing vegetation to recover and improving soil health (Schweitzer 2022). Projects focused on holistic management have also been developed, considering the overall impact of livestock on the ecosystem and aiming to balance meat production with soil health, biodiversity, and other environmental factors (Caro and Puch 2021). Finally, in the area of sustainable agriculture, projects promoting soil cover and no-till farming—where seeds are sown without prior soil preparation—are noteworthy (Tamayo et al. 2021). This practice helps conserve moisture, reduce erosion, and promote biological activity in the soil. Likewise, agroforestry projects that combine trees or shrubs with crops or livestock in the same area to enhance biodiversity, provide shade, and contribute to soil health have increased in the country.

6.6.2 Colombia

6.6.2.1 Regulatory Framework

According to the National Climate Change System (SISCLIMA 2023), which establishes the regulatory frameworks for green finance, the Central Bank of Colombia has led the creation of the National Sustainable Finance Policy, which aims to integrate environmental and social considerations into the financial system. The Financial Superintendence of Colombia has issued guidelines and standards to establish procedures for assessing and managing risks related to environmental and social aspects. Colombia has participated in international initiatives related to sustainable finance, such as the United Nations Principles for Responsible Banking (UNEP FI) and the United Nations Environment Program Finance Initiative (UNEP FI). These initiatives seek to align financial practices with sustainability goals and climate change mitigation. Colombian regulations have encouraged the inclusion of environmental, social, and governance (ESG) criteria in pension fund management. This involves considering sustainability factors in pension fund investment decisions.

6.6.2.2 Main Green Bond Issuers and ESG Stock Indices

Regarding corporate performance, three organizations in Colombia are particularly noteworthy. First, the Colombian bank Bancolombia has issued green bonds as part of its efforts to finance sustainable projects, such as renewable energy and energy efficiency. In 2016, the institution became the first financial entity in Latin America to issue green bonds for approximately $115 million, which were fully acquired by the International Finance Corporation (IFC), part of the World Bank Group, recognized as a leading global promoter of green investment (Bancolombia 2017). On the other hand, the state-owned oil company Ecopetrol has expressed interest in issuing green bonds as part of its initiatives to reduce its environmental footprint and promote more sustainable practices in the energy sector (Murcia, 2021). In 2021, the company successfully completed its first external public debt management operation for US$2 billion through a bond issuance in the international capital market to mitigate its environmental impact (Ecopetrol 2021). Finally, ISA (Interconexión Eléctrica S.A.), a company dedicated to electricity transmission, has also explored issuing green bonds for about $80 million to finance two projects on Colombia's northern coast that will enable the connection of non-conventional renewable energy sources (wind and solar) to the National Interconnected System (SIN) (ISA 2020).

The website of the Colombian Stock Exchange indicates that the COLIR is the only Colombian index with environmental, social, and governance (ESG) classification and captures social, environmental, or corporate governance initiatives. The COLIR index of the Colombian Stock Exchange lists the shares of 23 companies in the ESG category (Environmental, Social, and Governance) and is composed of (BVC 2023): GRUPO ÉXITO S.A., BANCO DAVIVIENDA S.A., BANCO BOGOTÁ S.A., BANCOLOMBIA S.A., BBVA COLOMBIA S.A., BOLSA DE VALORES DE COLOMBIA, CANACOL ENERGY LTD, CELSIA S.A. E.S.P., CEMETOS ARGOS S.A., CONSTRUCCIÓN DEL CÓNDOR S.A., CONSTRUCTORA CONCRETO S.A., CORFICOLOMBIANA S.A., ECOPETROL S.A., ETB S.A. E.S.P., GEB S.A. E.S.P., GRUPO RGOS S.A., GRUPO AVAL S.A., GRUPO NUTRESA S.A., GRUPO SURA S.A., INTERCONEXIÓN ELÉCTRICA S.A. E.S.P., MINEROS S.A., ORGANIZACIÓN TERPEL S.A., PROMIGAS S.A. E.S.P.

6.6.2.3 Projects for Green Finance

Colombia has also promoted the development of sustainable projects in the energy sector. For example, La Guajira, a region in northern Colombia, has been the site of several wind farm projects aimed at harnessing wind energy. These projects seek to diversify the country's energy matrix and reduce dependence on non-renewable sources (Hoyos et al. 2019). Similarly, the energy company Celsia developed the Toluviejo solar plant in the department of Sucre. This plant contributes to the generation of clean and sustainable energy by capturing solar radiation (EFIGIE 2023). The Ituango Project in Antioquia also stands out for its significance in renewable energy generation. Despite the challenges and delays faced during its construction, Ituango is expected to play a significant role in the country's hydroelectric capacity (Ochoa et al. 2017). The Biomass Plant in Palmira is a facility that generates energy from oil palm waste in Palmira. These projects aim to utilize agricultural residues to generate energy sustainably and strengthen sustainable finance.

6.6.3 Chile

6.6.3.1 Regulatory Framework

As of 2023, and according to Aguiló (2023), Chile has several regulatory frameworks and regulations that support green finance and sustainability. These regulations aim to promote the development of sustainable projects and investment in activities that generate environmental and social benefits. Law No. 20,780 on Income Tax establishes tax incentives for non-conventional renewable energy (NCRE) projects. Qualified projects can access tax benefits, which has contributed to the development of the renewable energy industry in Chile. In addition, the Chilean Social Responsibility Standard (NCh 3262) provides guidelines for implementing corporate social responsibility management systems. While it does not focus specifically on green finance, it highlights companies' commitment to sustainable and responsible practices. There is also the Financial Market Commission (CMF), which is the regulatory body for the financial market in Chile. The CMF has been promoting the integration of environmental, social, and governance (ESG) criteria in investment decisions. The CMF has also established regulations for the disclosure of ESG-related information by entities under its supervision. Additionally, Law 21,132 on Energy Efficiency aims to promote energy efficiency across various sectors, including incentives for the implementation of projects and measures that encourage the sustainable use of energy. Finally, the regulations of the Superintendence of Pensions (SP) have established guidelines for pension fund investment in sustainable financial instruments. This includes directives on the integration of ESG factors in investment decision-making.

6.6.3.2 Main Green Bond Issuers and ESG Stock Indices

Some of the organizations that have increased green bond issuance in Chile include the following: Enel Chile, a major company in the energy sector, has been active in the development of renewable and sustainable energy projects. For example, in 2020, the Enel Group published a Sustainability-Linked Financing (SLF) Framework, which sets out a series of key performance indicators (KPIs), targets, and principles to guide the Group's use of sustainable financial instruments in a comprehensive manner (Enel 2020). Enel is the first company to define such a framework for issuance through multiple financing solutions. Another major energy company, Colbún, began issuing green bonds in 2021 for US$600 million to finance energy projects; this issuance complies with the standards and requirements set by the Green Bond Principles of the International Capital Markets Association (ICMA) to qualify as a green bond (Colbún 2021). Similarly, AES Gener in 2021 issued the first green hybrid bond in the Americas for US$450 million in international markets, with a maturity of 60 years (to 2079) and an annual interest rate of 6.35%. This bond will be used to repurchase bonds maturing in 2021 and 2025 and to finance the purchase of the Los Cururos wind farm (110 MW wind power in the Coquimbo region) and the La Cebada substation (CMFS 2023). Finally, Chilean financial institutions have also increased their participation in green finance. In 2023, Banco de Chile completed the issuance of its first bond abroad under its new Sustainable Financing Framework (ESG Framework), with proceeds to be used to finance women-led enterprises. In 2022, Banco de Crédito e Inversiones (BCI) of Chile issued a green bond in the European market for an amount in Swiss francs equivalent to approximately US$216 million, with a term of 5.25 years and a placement rate of 0.5994%. With this transaction, BCI became the first Chilean bank to issue a public green bond outside the country, which was rated A2 by Moody's and A- by S&P (Ministry of Finance 2023).

In January 2021, the Chilean ESG stock index began operations under the name S&P IPSA ESG TILTED. This index aims to measure the performance of shares issued by companies that meet sustainability requirements and seek to improve their S&P DJI ESG score, which includes environmental, social, and corporate governance aspects. The S&P Dow Jones Indices website shows that this index is currently composed of 26 companies (S&P Global 2023): Enel Américas, Empresas CMCP S.A., Banco Santander de Chile, SQM, Banco de Chile, Empresas Copec S.A., Falabella S.A., Cencosud S.A., BCI, Parque Arauco S.A., CAP S.A., Viña Concha y Toro S.A., Embotelladora Andina S.A. (Series B), Entel, Aguas Andinas S.A., CCU S.A., Compañía Sud Americana de Vapores S.A., Itaú Corpbanca S.A., Plaza S.A., Sonda S.A., Inversiones Aguas Metropolitanas S.A., ILC S.A., and Cencosud Shopping S.A., 24. Grupo Security S.A., SMU S.A., and Ripley Corp S.A.

6.6.3.3 Green Finance Projects

Currently, Chile has seen a growing development of sustainable projects that promote green finance, particularly in the energy sector. Some of the main projects as of 2023 include, for example, the Cerro Tigre Wind Farm. This wind farm, located in the Antofagasta Region, is one of the largest wind energy projects in Chile. The facility makes a significant contribution to the country's clean energy generation capacity (Izquierdo 2015). The El Arrayán Wind Farm, located in the Coquimbo Region, is another major wind farm in Chile with substantial capacity. The Salvador Solar Plant, a photovoltaic solar plant, is located in the Atacama Region. Salvador is part of a complex of solar projects that take advantage of the high solar radiation in northern Chile. Similarly, the Pampa Norte Photovoltaic Plant, located in the Antofagasta Region, is a solar photovoltaic plant that is part of efforts to increase the country's solar energy generation capacity (Pérez 2020). Additionally, the Rucatayo Hydroelectric Plant, situated in the Los Ríos Region, is a hydroelectric facility that utilizes water resources for renewable energy generation. Finally, the Rubí Solar Park and the La Huella Photovoltaic Solar Power Plant are other facilities that have encouraged green investments in the country.

6.6.4 Brazil

6.6.4.1 Regulatory Framework

Regarding the regulatory framework for green finance in Brazil, the following guidelines can be mentioned (Marques et al. 2020): From the Central Bank of Brazil (Bacen), there is Resolution Bacen 4,327/2014. This resolution establishes rules for the implementation of social and environmental responsibility policies by financial institutions. Institutions are expected to consider environmental and social factors in their credit decisions and risk management. Similarly, Resolution Bacen 4,480/2016 aims to encourage the participation of financial institutions in sustainability-related activities, including investments in assets related to renewable energy. Additionally, through the Brazilian Securities and Exchange Commission (CVM), there is CVM Instruction 588/2017. This instruction regulates the creation and operation of green FIDCs (Receivables Investment Funds). Green FIDCs are investment funds that allocate their resources to the acquisition of receivables linked to green projects. On the other hand, the National Bank for Economic and Social Development (BNDES) has formulated the BNDES Social and Environmental Responsibility Policy. As a key entity in project financing in Brazil, BNDES has established social and environmental responsibility policies that include criteria for financing sustainable projects. This is complemented by the BNDES Climate Governance Program. BNDES has also developed specific initiatives to finance projects that contribute to climate change mitigation and adaptation. Finally, regarding the Capital Market, the role of B3 (Brasil, Bolsa, Balcão) can be highlighted. The Brazilian stock exchange B3 has introduced initiatives to promote transparency and the disclosure of sustainability-related information by listed companies.

6.6.4.2 Main Green Bond Issuers and ESG Stock Indices

Regarding the main green bond issuers in Brazil, a significant number of organizations can be mentioned. However, in 2023, Brazil made significant progress with a direct issuance by the government. In 2023, the Brazilian Government, through the National Treasury, announced the results of an operation in the U.S. market that included the issuance of a new 7-year benchmark, called GLOBAL 2031 ESG. The GLOBAL 2031 ESG matures on March 18, 2031, and was issued for an amount of $2 billion, with an annual yield for investors of 6.50%, corresponding to a spread of 181.9 basis points above the benchmark U.S. Treasury bond—the lowest level for new issuances in nearly a decade. The sustainable bond has an annual interest coupon of 6.25%, payable on March 18 and September 18 each year. The issuance was made at a price of 98.572% of its face value (GFL 2023). This marks the first sustainable bond issued directly by the Brazilian government. Among other organizations, Petrobras stands out. The state oil company Petrobras has explored the issuance of green bonds to finance projects related to the transition to cleaner and more sustainable energy sources; between 2015 and 2022, the company reduced its CO_2 emissions by 39% and issued a green bond worth $1.25 billion in 2022 (Barbosa et al. 2023). The National Bank for Economic and Social Development (BNDES) in 2017 was the first bank to issue green bonds in the international capital market, and the first financial institution to issue green financial notes in the local market in 2020. Another notable case is the chemical company Braskem, which has issued a significant amount of green bonds maturing in 2032 with a return of 15.4% (Braskem 2023). Finally, the mining company Vale can also be mentioned, which has issued green bonds to finance projects that contribute to sustainability and reduce the environmental impact of its operations, as well as the company Natura & Co cosmetics, which includes brands such as Natura, Avon, and The Body Shop, has been involved in sustainability initiatives and may consider issuing green bonds in the short and medium term.

On August 31, 2020, the São Paulo Stock Exchange (BOVESCA) launched the S&P/B3 Brazil ESG Index. This index takes into account the financial and economic situation of companies, corporate governance aspects, and environmental and social standards. The objective of this index is to serve as a broad and representative index of the Brazilian stock market with an enhanced ESG profile. Each company has a specific weight in the index according to its ESG score. Currently, the index is composed of 47 companies (BOVESCA 2023): AE-RIS, CLEARSALE, TIENDAS RENNER, SABESP, ALPARCATAS, COPEL, RE-VISTA LUIZA, SANTANDER BR, AMBEVS/A, DASA, MATER DEI'S, BAP SANTOS, AMERICANO, BR GETNET, ODONTOPREVIO, MELIUZ, CO AREZZO, HIPERA, TAESA, TEC POSITIVO, B3, INTELBRAS, PACA MENOS, NEOGRID, BANCO PAN, IRANi, ITAUUNIBANCO, TELEF BRASIL, BESAR, JALLESMACHAD, PLANPLAN, QUEREP, BRADESCO, KEPLER, LOCAWEB, QUALICORPO, MODAS CEA, LOCALIZA, SINQIES, ACRICOLA SLC, CESP, VIA, WIZ S. A., TOTV, CIELO, SUZANO S. A., and MOSAICO.

6.6.4.3 Green Finance Projects

In recent years, Brazil has developed significant projects aimed at the growth of green finance. Sustainable mobility projects, in particular, have gained prominence in the country. For example, the Bus Rapid Transit (BRT) system in Curitiba uses dedicated bus lanes, prepaid stations, and high-capacity vehicles to provide fast and efficient transportation, in addition to the implementation of electronic ticketing systems, which allow for more efficient access to public transport, streamlining the payment process and reducing reliance on paper tickets. In terms of environmental projects, several incentives have also been implemented to promote the use of alternative fuels and more sustainable vehicles. These incentives aim to encourage the adoption of cleaner technologies, reduce greenhouse gas emissions, and advance toward a more diversified and sustainable energy matrix. For example, the Inovar-Auto program and the Rota 2030 program, a government initiative that incentivizes the manufacture of more efficient and less polluting vehicles. It sets efficiency targets for automakers and provides tax benefits to those who meet these targets (Claro and Bruno 2020). Also worth mentioning is Compressed Natural Gas (CNG). CNG is promoted as a cleaner alternative to conventional fuel. Natural gas vehicles may benefit from tax incentives and vehicle tax reductions in some jurisdictions. Similarly, the Proálcool (National Alcohol Program) is an initiative to promote sustainable investment. This historic program has encouraged the use of ethanol as a vehicle fuel, providing incentives for the production and use of ethanol, including tax benefits and reductions. Finally, in the energy sector, Brazil is a leader with projects such as the Lagoa do Barro Wind Farm (Piauí) and the Lagoa do Ouro Wind Complex (Pernambuco). In the solar sector, there are projects such as the Pirapora Solar Park (Minas Gerais) and the Nova Olinda Solar Park (Piauí). Brazil is particularly known for its use of biomass and biogas energy, with projects like the Barcarena Biomass Plant (Pará) and the Bandeirantes Landfill Biogas Project (São Paulo) serving as benchmarks for this type of investment (Xu et al. 2022). Notably, at the Latin American level, is the Tidal Energy Project in Pecém (Ceará). Brazil has explored pilot projects to harness tidal energy along its coast; although still in early stages of development, these efforts demonstrate a commitment to advancing a sustainable energy transition.

6.6.5 Mexico

6.6.5.1 Regulatory Framework

The regulatory frameworks in Mexico that can foster green finance are diverse depending on the focus. According to BANXICO (2015), first to note is the Securities Market Law (LMV). The LMV has been amended to include provisions related to the issuance of sustainable securities and the disclosure of environmental, social, and corporate governance (ESG) information by listed companies. In the energy sector, there are various laws, such as the Energy Transition Law, which establishes measures to promote the transition to a more sustainable and low-carbon energy matrix. It contains provisions to promote energy efficiency, the use of renewable energy, and the reduction of greenhouse gas emissions. The Law for the Use of Renewable Energy and Financing of the Energy Transition aims to encourage energy generation from renewable sources and establishes mechanisms to finance energy transition projects. The Energy Planning Law sets out the principles and criteria for planning the energy sector, including the promotion of renewable energy and energy efficiency. Finally, the Bioenergy Law regulates the production and use of bioenergy, with the goal of promoting renewable and sustainable energy sources. These types of regulations open the door to green finance, which would be complemented by CNBV and Banxico standards. The National Banking and Securities Commission (CNBV) and the Bank of Mexico (Banxico) have issued regulations and guidelines related to the integration of environmental, social, and corporate governance (ESG) criteria in financial operations. As a result, Mexico presents an adequate regulatory landscape for the promotion of green finance, especially in the energy sector.

6.6.5.2 Main Green Bond Issuers and ESG Stock Indices

Regarding green bond issuance, the corporate sector in Mexico has played an active role. For example, Grupo Bimbo, one of the largest baking companies in the world, has participated in green bond issuances to finance projects related to energy efficiency and carbon emission reduction. In particular, this business group in 2023 raised 15 billion pesos with its first sustainability-linked bond issuance on the Mexican Stock Exchange (BMV), with a fixed annual interest rate of 9.24% (Santiago 2023). Another important example is Fibra Uno, a real estate investment trust, which in 2021 issued sustainable bonds for approximately 400 million dollars on the BMV. The Fibra Uno issuances, with ticker symbols FUNO21X and FUNO21-2X, have maturities of 7 and 4 years, respectively, and are aimed at projects to reduce energy consumption or improve resource efficiency; expenses related to the construction, development, acquisition, maintenance, and operation of clean energy projects such as solar, wind, and low-impact hydroelectric, as well as investments in socioeconomic advancement and empowerment (GFL 2021). Cemex, a global building materials company, has been involved in green bond issuances to finance projects that contribute to environmental sustainability, such as reducing carbon emissions in its operations. The cement company aims to have 85% of its debt linked to sustainability by 2030, and in 2023 issued nearly 350 million dollars in long-term sustainability-linked notes toward that goal (Expansión 2023). Meanwhile, Grupo Aeroportuario del Pacífico (GAP), which operates several airports in Mexico, in 2023 raised approximately 315 million dollars through the issuance of sustainability-linked bonds, with which it commits to achieving targets to reduce the carbon footprint of its operations (Gonzalez 2023). In the public sector, a notable case is Nacional Financiera (NAFIN), a development finance institution in Mexico, which has been involved in green bond issuances to finance sustainable projects in sectors such as renewable energy and energy efficiency. In particular, in 2016, it issued the first Mexican green bond for 500 million dollars, making it the first development bank in Latin America to issue a green bond and the first green bond in Latin America to receive a recognized international sustainability certification (NAFIN 2020). A year later, in 2017, the National Bank of Public Works and Services (Banobras) designed the Sustainable Funding Strategy, aimed at financing green and social infrastructure, particularly through green bonds with a gender perspective. This demonstrates the active participation of the public sector in Mexico in green bond issuance.

In 2011, the Mexican Stock Exchange launched the Sustainable Price and Quotation Index (IPC Sustentable), which was renamed on June 21, 2020, as the S&P/BMV Total Mexico ESG Index. This index uses stock selection criteria based on environmental, social, and corporate governance principles. According to the Mexican Stock Exchange website, it currently comprises 30 companies (BMV 2023): Arca Continental S.A.B. de C.V., Grupo Rotoplas S.A.B. de C.V., Alpek S.A.B. de C.V., Alsea S.A.B. de C.V., América Móvil S.A.B. de C.V., Axtel S.A.B. de C.V., Grupo Bimbo S.A.B. de C.V., Grupo Bolsa Mexicana de Valores S.A.B. de C.V., CEMEX CPO S.A.B. de C.V., Grupo Elektra S.A.B. de C.V., Fomento Económico Mexicano S.A.B. de C.V., Prologis Property México S.A. de C.V., Fibra Uno S.A. de C.V., Grupo Aeroportuario del Pacífico S.A.B. de C.V., Grupo Financiero Banorte S.A.B. de C.V., Grupo México S.A.B. de C.V., Grupo México Transportes S.A.B. de C.V., Kimberly Clark de México S.A.B. de C.V., Coca Cola Femsa S.A.B. de C.V., Genoma Lab International S.A.B. de C.V., El Puerto de Liverpool S.A.B. de C.V., Orbia Advanced Corporation S.A.B. de C.V., Qualitas Controladora S.A.B. de C.V., Regional S.A.B. de C.V., PLA Administradora Industrial S.R.L., Grupo Televisa S.A.B., Grupo Traxion S.A.B. de C.V., Corporación Inmobiliaria Vesta S.A.B. de C.V., Controladora Vuela CIA de Aviación S.A.B. de C.V., and Wal-Mart de México S.A.B. de C.V.

6.6.5.3 Projects for Green Finance

As of 2024, various projects related to green bond issuance are being developed in Mexico, particularly in the energy sector. Notable examples include the Reynosa I and II Wind Farms, located in Tamaulipas, which are part of Mexico's efforts to increase wind power generation capacity. Additionally, the Villanueva I and II Solar Farms, located in the state of Coahuila, are part of the solar energy infrastructure aimed at generating green investments, along with the Don José Solar Park in the state of Guanajuato and the Durango Photovoltaic Project. Furthermore, Mexico has developed projects that have encouraged sustainable investment, such as the Maya Train, which aims to improve connectivity in the southeastern region of Mexico and has been mentioned in the context of green bond issuance. Likewise, transportation infrastructure has benefited from green bond issuance, with reported projects for the installation and repair of equipment to restore the quality of passenger transport services, as well as the acquisition of new electric vehicles (GIZ 2020). Finally, there are currently initiatives in Mexico that promote green investments in protected agriculture projects, forestry projects, commercial consulting, and management to enhance the social and economic development of rural women, as well as loans to promote the financial inclusion of women in order to contribute to gender equality, under the responsibility of the Trust Funds for Rural Development (FIRA) (BANXICO 2021).

6.7 Conclusions

This chapter has explored the landscape of green finance in Latin America by analyzing both the general context and the specific situation of the five countries with the highest GDP in the region. The review has highlighted several important aspects that deserve special attention. First, it is clear that green finance holds tremendous transformative potential to drive sustainable development in Latin America. Through innovative financial instruments, investments can be channeled into projects that not only generate economic returns but also promote environmental conservation and mitigate the impacts of climate change. This integrated approach not only responds to the growing demands of socially conscious investors but also lays the groundwork for more equitable and sustainable economic growth. On the other hand, the need for a robust and coherent regulatory framework to support the development of green finance in the region has become evident. Collaboration among governments, financial institutions, and civil society is a crucial element to ensure the effective implementation of policies that foster responsible investment and the adoption of sustainable financial practices. Transparency and accountability must be fundamental pillars in the management of these resources, thereby ensuring that the benefits are distributed equitably. Likewise, the role of education and awareness should not be underestimated. A deep understanding of green finance among the various stakeholders in the region can accelerate the transition toward a more sustainable economic model. Educational campaigns, training, and outreach are powerful tools to engage society at large and economic agents in particular in the adoption of responsible financial practices.

A review of the status of the selected countries shows that Chile has emerged as a regional leader in green finance, consolidating its position through the implementation of policies and regulations that encourage sustainable investment. The issuance of green bonds and the promotion of renewable energy projects have helped position the country as a benchmark in the region. Argentina, for its part, has experienced progressive growth in environmental awareness and the adoption of sustainable financial practices. However, the lack of specific regulatory frameworks and the need to strengthen infrastructure for green projects remain challenges to be addressed. Colombia has made progress in incorporating environmental criteria into its financial policies, promoting investment in sectors such as sustainable agriculture and forest conservation. Despite this, greater efforts are needed to fully align the financial system with sustainability principles. Brazil, as one of the largest economies in the region, faces tensions between economic expansion and environmental preservation. Although there has been an increase in the issuance of green bonds and investment in renewable energy, deforestation, and pressure on natural resources pose significant challenges. Finally, Mexico has advanced in integrating sustainable financial practices, standing out in the issuance of green bonds and the promotion of clean energy projects. Despite these achievements, the need to strengthen environmental oversight and regulation, as well as to diversify green investments, are crucial issues to address.

In summary, the countries analyzed in the region show a positive trend toward the adoption of green finance, but challenges remain in terms of regulation, infrastructure, and balancing economic development with environmental sustainability. Regional collaboration and ongoing commitment to environmental and social policies may be key to further advancing the transition toward a more sustainable financial system in the region, gradually integrating a greater number of economies into green bond issuance.

References

Aguiló J (2023) Analysis of the institutional and regulatory progress of green finance in Chile. Revista de Derecho Económico 80(1):55–72. https://doi.org/10.5354/0719-7462.2023.70588

Bancolombia (2017) Bancolombia recognized as the first private bank in Latin America to issue green bonds. Retrieved from: https://shorturl.at/amoT5

BANXICO (2015) International and national principles. Retrieved from: https://www.banxico.org.mx/sistema-financiero/principios-financiamiento-v.html. Accessed 01–11–2024

BANXICO (2021) Issuance of thematic bonds by development banks and promotional institutions. Retrieved from: https://shorturl.at/dlCG5. Accessed 01–31–2024

Barbosa F, Hyland S, and Knoll T (2023) The key role of Petrobras in Brazil's energy transition. Retrieved from https://shorturl.at/ahoX8. Accessed 01–14–2024

BMV (2023) Mexican stock exchange. Retrieved from: https://www.bmv.com.mx/es/empresas-listadas. Accessed 02/11/2024

BOVESCA (2023) Brazilian stock exchange. Retrieved from: https://www.b3.com.br/pt_br/para-voce. Accessed 02/11/2024

BVA (2023) Argentine stock exchange. Retrieved from: https://www.byma.com.ar/emisoras/indice-de-sustentabilidad-esp/. Accessed: 02/11/2024

BVC (2023) Colombian stock exchange. Retrieved from: https://www.bvc.com.co/indices-de-sostenibilidad. Accessed 02/11/2024

Braskem (2023) 2022 ESG sustainability annual report. Retrieved from: https://shorturl.at/qswQZ. Accessed 01/16/2024

Caro V, Puch F (2021) Sustainable finance: not a trend, but a duty. Ibero-American Institute of Securities Markets. Retrieved from: https://shorturl.at/fwPT9. Accessed 01/02/2024

Castro E (2007) Environment issues in the Latin American region. Department of Economic and Social Affairs, United Nations. Retrieved from: https://www.un.org/esa/socdev/unpfii/documents/workshop_IPPE_castro.pdf. Accessed 01/02/2024

Chancay-Cedeño A (2016) Corporate social responsibility: history and challenges for its study. FIPCAEC 1(1):39–50. https://doi.org/10.23857/fipcaec.v1i1.10

Claro J, Bruno L (2020) State and the automotive industry in Brazil: analysis of the Inovar-Auto and Route 2030 policies. Entrelugar 11(21):101–127. https://doi.org/10.30612/el.v11i21.12051

CMFS (2023) Congratulations to AES Argentina for issuing green bonds worth US$48.4M! Retrieved from: https://cmfs.org.mx/noticias/felicidades-a-aes-argentina-que-emitio-bonos-verdes-por-us48-4m/. Accessed 01/11/2024

Colbún (2021) Colbún issues green bond for US$600 million to finance energy projects. Retrieved from: https://www.colbun.cl/corporativo/sala-de-prensa/noticias-y-comunicados/detalle/colbun-coloca-bono-verde-por-us-600-millones-para-financiar-proyectos-de-energia. Accessed 01/08/2024

Contreras O, Pedraza A, Martínez M (2017) Impact investing as a means to promote sustainable development: a multi-case approach at the company level in Colombia. Estudios Gerenciales 33(142):13–23. https://doi.org/10.1016/j.estger.2017.02.002

Correa F (2007) Economic growth, social inequality, and the environment: empirical evidence for Latin America. Revista Ingenierías Universidad De Medellín 6(10):12–30

De Santiago J (2020) Socially responsible investment: sustainable funds and their profitability. Bachelor's Thesis. Universidad Pontificia Comillas

Dessai S, Lacasta N, Vincent K (2003) International political history of the Kyoto protocol: from the Hague to marrakech and beyond. Int Rev Environ Strateg 4(2):183–205

ECLAC (2023) Latin American economic outlook 2022: towards a green and just transition. Retrieved from: https://hdl.handle.net/11362/48426. Accessed 01/11/2024

Ecopetrol (2021) Ecopetrol successfully completes its first external public debt management operation for US$2 billion through bond issuance in the international capital market. Retrieved from: https://rb.gy/7hev1y. Accessed 01–05–2024

EFIGIE (2023) Irish company arrives in Colombia to build its first solar park in sucre. Retrieved from: https://rb.gy/nkz7ni. Accessed 01–04–2024

Elbers J (2013) The Copenhagen climate summit versus real climate change. Letras Verdes 6:16–18. https://doi.org/10.17141/letrasverdes.6.2010.874

Enel (2020) Enel launches a sustainability-linked financing framework. Retrieved from: https://rb.gy/t1lskm. Accessed 01–07–2024

Ferrer E (2013) Green bonds: incentive regime for environmentally conscious companies. Pensar En Derecho 3(2):87–100

Gafoor A, Perumbalath S, Daimari P, Naheem K (2024) Trends and Patterns in green finance research: a bibliometric study author links open overlay panel. Innov Green Dev 3(2):100119. https://doi.org/10.1016/j.igd.2023.100119

Gallardo D (2006) Commitment to sustainable development: the equator principles. Cuadernos De Economía 24(45):207–221

Garayoa F, García A (2018) Green Bonds and social bonds as drivers of change. Boletin De Estudios Economicos 73(224):233–250

GFL (2023) Brazil announces the result of its first sustainable bond issuance. Retrieved from: https://greenfinancelac.org/resources/news/brazil-announces-the-result-of-its-first-sustainable-bond-issuance/. Accessed 01–12–2024

GIZ (2020) Mexico City green bond. Retrieved from: https://iki-alliance.mx/wp-content/uploads/12-CDMX.-Bono-Verde.pdf. Accessed 01–31–2024

Guerrero S (2021) State promotion for the development of green finance. Retrieved from: https://opus.bibliothek.uni-wuerzburg.de/files/32359/978-3-95826-149-5_GuerreroSabogal_OPUS_32359.pdf. Accessed 12–22–2023

Herrera, H. (2023) Green finance under the Escazú agreement. Heinrich-Böll-Stiftung. Retrieved from: https://mx.boell.org/es/2023/08/07/finanzas-verdes-bajo-el-acuerdo-de-escazu. Accessed 01–03–2024

Hoyos B, Velez F, Morales D (2019) Wind energy and territory: geographic information systems and multicriteria decision methods in La Guajira (Colombia). Ambiente y Desarrollo 23(44):1–20. https://doi.org/10.11144/Javeriana.ayd23-44.eets

Hwang Y (2023) The synergy effect through the combination of the digital economy and the transition to renewable energy on green economic growth: empirical study of 18 Latin American and Caribbean Countries. J Clean Prod 418:138146. https://doi.org/10.1016/j.jclepro.2023.138146

Inderst G, Stewart F (2018) Incorporating environmental, social and governance (ESG) factors into fixed income investment. World Bank Group. Retrieved from: https://documents1.worldbank.org/curated/en/913961524150628959/pdf/125442-REPL-PUBLIC-Incorporating-ESGFactors-into-Fixed-Income-Investment-Final-April26-LowRes.pdf. Accessed: 06-02-2024

ISA (2020) Successful issuance of green bonds by ISA: the first by a real sector company in the public securities market. Retrieved from: https://rb.gy/hnrd7d. Accessed 01–06–2024

Izquierdo A (2015) Study of solutions for the elimination of the impact on the watercourse network at the future site of the Cerro Tigre wind farm (Antofagasta, Chile). Undergraduate thesis. Polytechnic University of Valencia

Keeble B (2007) The Brundtland report: 'Our common future.' Medicine and War 4:17–25. https://doi.org/10.1080/07488008808408783

La Torre M, Leo S, Palma A, Salazar J (2024) Public spending and green finance: a systematic literature review. Res Int Bus Financ 68:102197. https://doi.org/10.1016/j.ribaf.2023.102197

Liu Y, Zhao C, Dong K, Wang K, Sun L (2023) How does green finance achieve urban carbon unlocking? Evidence from China. Urban Climate 52:101742. https://doi.org/10.1016/j.uclim.2023.101742

Longo L, Tomasini D, Henry A (2022) The role of finance in a world facing climate change. the argentine opportunity. Ecogralia 6(12):17–21

Marques F, Junqueira G, Medeiros M, Ferrando T (2020) Green finance and the dismantling of environmental regulation in service of the far right in Brazil. Ecología Política 59:57–66

McCauley D, Pettigrew K, Heffron R, Droubi S (2023) Identifying, improving, and investing in national commitments to just transition: Reflections from Latin America and the Caribbean. Environ Sustain Indicators 17:100225. https://doi.org/10.1016/j.indic.2023.100225

Mendoza M, Greenleaf M, Thomas E (2021) Green distributive politics: legitimizing green capitalism and environmental protection in Latin America. Geoforum 126:1–12. https://doi.org/10.1016/j.geoforum.2021.07.012

Ministry of Economy (2021) Regulations on Sustainable Finance. Retrieved from: https://www.argentina.gob.ar/cnv/finanzas-sostenibles/normativa. Accessed 01–15–2024

Ministry of Finance (2023) Ministry of Finance publishes the 2019 Green Bonds Report. Retrieved from: https://www.hacienda.cl/areas-de-trabajo/finanzas-internacionales/oficina-de-la-deuda-publica/bonos-sostenibles/bonos-verdes. Accessed 01–10–2024

Municipality of Mendoza (2023) The City, together with Banco Galicia, launches the Green Bond to combat climate change. Retrieved from: https://rb.gy/b1ffur. Accessed 01–12–2024

Murcia A (2021) Issuance of green bonds for a sustainable economy in Colombia. Retrieved from: https://www.uexternado.edu.co/wp-content/uploads/2021/12/Emisio%CC%81n-de-Bonos-Verdes-para-una-Economi%CC%81a-sostenible-en-Colombia.pdf. Accessed: 12–27–2023

Ochoa V, Marín W, Osejo A (2017) Valuation of ecosystem services in the area of influence of the Ituango Hydroelectric Project—Antioquia. Retrieved from: http://tinyurl.com/3fysef2h. Accessed 01–06–2024

Ospina H, Marrugo L, Castilla L, Castellón L, Cantillo A, Bolivar L, Salas K, Zamora R (2023) Analysis in circular economy research in Latin America: a bibliometric review. Heliyon 9(9):e19999. https://doi.org/10.1016/j.heliyon.2023.e19999

Pérez C (2020) Regulations and public policies aimed at the growth of renewable energy in Chile. Revista De Derecho Ambiental 14:9–41. https://doi.org/10.5354/0719-4633.2020.54038

Samaniego J, Schneider H (2023) Fifth report on climate finance in Latin America and the Caribbean, 2013–2020. Economic Commission for Latin America and the Caribbean (ECLAC). Retrieved from: http://repositorio.cepal.org/handle/11362/49085. Accessed 01–05–2024

Schweitzer A (2022) The new political-ecological regime and the consensus of uneven and combined transition in the Global South. the case of Argentina. Ecología Política 64:36–43

SISCLIMA (2023) National action framework. Retrieved from: https://finanzasdelclima.dnp.gov.co/financiamientoclimatico/generalidades/marco-regulatorio/Paginas/marco-regulatorio-nacional.aspx. Accessed 01–31–2024.

Stanley L (2021) Green finance in Latin America and the Caribbean: debates, weaknesses, challenges, and threats. Fundación Carolina. Retrieved from: http://repositorio.cedes.org/handle/123456789/4665. Accessed: 01–05–2024

S&P Global (2023) S&P500 ESG index. Retrieved from: https://www.spglobal.com/spdji/es/indices/esg/sp-500-esg-tilted-light-index/#overview. Accessed 02–11–2024

Tamayo G, Merchán M, Gómez M (2021) New challenges for public and private administrative management: social responsibility and sustainable finance. VinculaTégica, 7(1):856–867. https://doi.org/10.29105/vtga7.2-17

Trina J (2019) Use of green bonds for financing bioenergy projects in Argentina. Undergraduate Thesis. National University of Rosario

Veltmeyer H (2022) Extractivism and beyond: Latin America debates. Extract Indus Soc 11:101132. https://doi.org/10.1016/j.exis.2022.101132

Viñuales V, Pueyo R (2002) Socially responsible investment. Fundación Ecología y Desarrollo: Zaragoza

Wu G, Liu X, Cai Y (2024) The impact of green finance on carbon emission efficiency. Heliyon 10(1):e23803. https://doi.org/10.1016/j.heliyon.2023.e23803

Xu J, Liu Q, Wider W, Zhang S, Ashraf M, Jiang L, Naces L, An Z (2024) Research landscape of energy transition and green finance: a bibliometric analysis. Heliyon in Press. https://doi.org/10.1016/j.heliyon.2024.e24783

Xu D, Salem S, Awosusi A, Abdurakhmanova G, Altuntas M, Oluwajana D, Kirikkaleli D, Ojekemi O (2022) Load capacity factor and financial globalization in Brazil: the role of renewable energy and urbanization. Front Environ Sci 9:823185. https://doi.org/10.3389/fenvs.2021.823185

Yi H, Hao L, Liu A, Zhang Z (2023) Green finance development and resource efficiency: a financial structure perspective. Resour Policy 85(Part A):104068. https://doi.org/10.1016/j.resourpol.2023.104068

Zhang G, Guo B, Lin J (2023) The impact of green finance on enterprise investment and financing. Financ Res Lett 58(Part D):104578. https://doi.org/10.1016/j.frl.2023.104578

Contents

Governance in the Health Sector and the SDGs in Mexico: Lessons from the COVID-19 Pandemic

7

Pilar Pérez-Hernández, Luis Manuel Castillo Chávez, and Hortensia Gomez Viquez

Abstract

The COVID-19 pandemic revealed the need to reconsider the variables that influence the achievement of the Sustainable Development Goals (SDGs), especially health governance. This study analyzes governance in Mexico's health sector in relation to SDG 3 (Good Health and Well-being), using a qualitative and documentary methodology based on literature on institutional capacities, health system characteristics, and available indicators. Significant limitations were identified: information is incomplete, heterogeneous, and not timely, which hinders decision-making. There are disparities between subsystems in terms of budget and coverage, as well as a lack of continuity due to changes in government every six years, which impedes collaboration among system actors. The Mexican Social Security Institute bears the greatest burden in the system, despite limited human resources and infrastructure. The lack of coordination, weak governance, and social determinants of health highlight the urgency of a more comprehensive approach. Despite these weaknesses, progress has been made in reducing maternal and infant mortality. The study concludes that strengthening governance is key to achieving health objectives.

Keywords

Governance · Health sector · Health policy · Mexico

7.1 Introduction

Humanity is currently facing the urgent need for global health governance. The implementation of the 2030 Agenda and the achievement of the Sustainable Development Goals (SDGs) require the identification and mobilization of multiple stakeholders, as well as multilevel coordination in which they collaborate to address the present and future needs of society. This is occurring in a context where the COVID-19 pandemic has affected the ability of stakeholders to coordinate their resources and direct their efforts toward achieving the SDGs.

The factors that impact health, the economy, and the environment are known as social determinants of health (SDH), such as the pace of societal growth, market globalization, increased mobility and insecurity, energy consumption and climate change, food security, concerns about risk and

P. Pérez-Hernández (✉) · L. M. C. Chávez · H. G. Viquez
Center for Economic, Administrative, and Social Research of the National Polytechnic Institute, Mexico City, Mexico
e-mail: mpperez@ipn.mx

L. M. C. Chávez
e-mail: lcastilloc2100@alumno.ipn.mx

H. G. Viquez
e-mail: hgomezv@ipn.mx

P. Pérez-Hernandez and R. Huerta Cuervo (eds.), *Social Well-Being and the Challenges in the Sustainable Development Goals*, Sustainable Development Goals Series,
https://doi.org/10.1007/978-3-032-14294-8_6

safety, and the reach of the media. Since 2013, the WHO has developed documents encouraging governments to modify or influence these SDH to change people's habits and promote positive health outcomes (Schoenfeld, 2019).

From this global perspective, the impact of phenomena such as extreme weather events, large-scale involuntary migration, epidemics, natural disasters, economic crises, terrorist attacks, or cyber fraud becomes particularly relevant, as these events, with a certain probability of occurrence, affect people's lives. These types of events, known as global risks, appear unexpectedly, can cause damage and affect the course of public action (García 2021).

The COVID-19 pandemic caused a massive collapse of human capital at critical stages of the life cycle, disrupting the developmental trajectory of millions of children and young people in low- and middle-income countries. More than 7 million lives were lost worldwide, and it is estimated that nearly 100 million people were pushed into extreme poverty (mainly in South Asia and Southern Africa), leading to a 14% drop in remittances. By the end of 2021, 40 million people who would have been employed under normal circumstances (without the pandemic) were not, exacerbating youth unemployment trends. Their incomes shrank by 15% in 2020 and 12% in 2021, while new entrants with lower educational attainment are expected to earn 13% less during their first 10 years in the labor market. Furthermore, today's students could lose up to 10% of their future earnings due to the educational crises triggered by COVID-19.

The current cognitive deficit among children could translate into a 25% reduction in their income as adults. For the first time in history, the global Human Development Index value declined, setting the world back to the period immediately following the adoption of the 2030 Agenda for Sustainable Development. During the first year of the pandemic, the global prevalence of depression and anxiety increased by more than 25%. Low-income individuals, especially those struggling to afford basic needs such as rent and food, suffered disproportionately across various countries (WHO 2024; Schady et al. 2023; UNDP 2022, 2023; Coneval 2023).

Despite this scenario, governments have not incorporated sufficiently robust institutional mechanisms into their strategies to ensure risk management, the adoption of measures to mitigate its effects, and the establishment of channels for citizen participation (García 2021; Schoenfeld 2019; Díaz-Castro et al. 2021).

In Mexico, COVID-19 resulted in 335,000 deaths (as of January 2024); in 2020, Gross Domestic Product (GDP) fell by 8.3%, although it grew by 4.8% in 2021, marking the steepest decline since the Great Depression of the 1930s. Before the pandemic, according to 2020 data, nearly 41.9% of the Mexican population was living in poverty (35.4% in moderate poverty and 8.5% in extreme poverty), and by 2022 this figure rose to 36.3% (29.3% in moderate poverty and 7.1% in extreme poverty).

In the last quarter of 2019, the informality rate was reported at 56.2%; by the end of 2020, nearly one million micro, small, and medium-sized enterprises (MSMEs) had closed (approximately 20.8%), and 86.6% of businesses were affected in some way by the pandemic. Regarding health care deprivation, in 2018 it was reported at 16.2%, and by 2022 it had risen to 39.1% of the population, representing more than 50 million people (UNDP 2022, 2023; Coneval 2023; WHO 2024).

The exact economic, social, and political effects of the COVID-19 pandemic have yet to be fully determined, although they are estimated to be the most severe since the end of World War II. Nevertheless, the socioeconomic impacts largely depend on which of the possible health scenarios materializes, as well as the timeliness and appropriateness of the economic policy responses implemented by each country's authorities.

In light of this situation, health requires a significant shift in the design and implementation of public policies, as addressing deficiencies necessitates the involvement of a range of stakeholders—a complexity that a mechanistic approach (focused on medical sciences) has not resolved (Hufty 2010). Research on governance involves identifying the political factors that limit efficiency, power relations, or the role of institutions in problem definition, and, once identified, seeking to resolve them. In this, the social sciences play an instrumental role in understanding obstacles and overcoming them.

Political factors are often perceived as obstacles to problem-solving rather than as an expression of social complexity. In complex societies, there are actors with conflicting interests; some even take advantage of tragic situations and primarily defend their own interests, which are not always economic but sometimes ideological. Moreover, they not only seek to gain ground but also attempt to influence the very definition of the rules of the game, and even if all stakeholders wish to resolve a situation, they may not succeed, and the sum of individually rational actions may result in a suboptimal outcome (Hufty 2010).

Governance in public health is an issue that must be addressed beyond technical considerations, where the social sciences can provide approaches, tools, and methodologies to decision-makers. This paper analyzes the extent to which governance capacities have been developed in Mexico's health sector to achieve the SDGs. To this end, the first section reviews the literature on governance capacities in general and in the health sector in particular; the next section characterizes the health sector, followed by a review of SDG 3 indicators (focused on primary and secondary care levels); the fourth section addresses the trials facing the health sector in Mexico, and finally, the conclusions are presented.

7.2 The Role of Governance Capacities in Contemporary Society

The word government comes from the Greek *kubernao* (to steer or direct), while governance is a concept that has gained popularity in political, social, economic, and ethical discourse. Although ambiguities persist regarding its core meaning, it quickly evolved into a new paradigm aimed at explaining the transformations of the State in recent decades (Santos 2022). According to Hufty (2010), the literature identifies four stages of governance:

a. *Governance in organizations.* It is observed that formal hierarchy does not necessarily correspond to effective hierarchy, and that relationships among employees are decisive in the organization of work. Likewise, the reduction of transaction costs motivates the creation of hierarchical organizations, but beyond a certain level of complexity (multiplicity of actors and units and increasing specialization), internal administrative costs rise. Corporate governance emerges when an organization becomes a space of cooperation and conflict among various types of actors, with their own norms, interests, practices, and coordination mechanisms.
b. *Global governance.* The aforementioned contributions resonated in the field of international relations. Coordination among States occurs horizontally, paving the way for the governance perspective and enabling the creation of agreements to address coordination problems beyond the State, thereby institutionalizing international life.
c. *Modern governance.* The debates of the 1970s regarding the role of the State in industrialized countries, the economic, political, and social crises of the 1970s and 1980s, as well as the perceived crisis of governability due to the postwar expansion of the state apparatus, led to multiple redundancies and inefficiencies. Additionally, globalization reduced the scope of public policy influence. Although the weakening of the State remains under discussion, several studies indicate that by focusing on its core functions, the State strengthened its relationship with key stakeholders.
d. *Good governance.* In their efforts to promote economic growth in late-industrializing countries, multilateral organizations such as the World Bank (WB), International Monetary Fund (IMF), and Inter-American Development Bank (IDB) deepened their intervention strategies upon recognizing that the success or failure of structural adjustment policies depended on non-governmental political actors. Consequently, they sought a methodological tool to identify the loci of effective power within political systems (both formal

and informal, state and non-state), and found the concept of governance (the manner in which power is exercised in the management of a country's economic and social resources for development). Governance thus shifted from being an analytical tool (*governance*) to a political instrument for societal transformation (*good governance*). Its use acquired a normative connotation, referring to what "should be" rather than what "is." Regardless of the meaning of these criteria (which, in fact, stem from the moral codes and political models of industrialized countries), when imposed externally, they can only be perceived as part of a power relationship.

The term has evolved and acquired multiple interpretations and definitions. Since its origin, but particularly over the past three decades, it has shared common elements: the involvement of government and civil society, with the aim of improving, increasing efficiency, collaborating, and cooperating to solve a common problem (Schoenfeld, 2019; Santos 2022). In this work, we adopt the definition of governance proposed by Hufty (2010) as "a type of social fact, the formal and informal collective processes of decision-making and the development of social norms in relation to public affairs."

In every society, whether local, national, or international, and at any point in its history, governance processes can be observed. It is not a phenomenon limited to our era. There are social norms (legal or customary) and institutions (recurrent systems of social norms) that make it possible to live together. Each society develops its own modes of governance, its systems for decision-making or conflict resolution among its members, its norms, and institutions (Hufty 2010).

Regarding the academic debate on governance in Latin America, it has been rather limited, and the dominant notion has been disseminated by international cooperation donors; these actors have played a central role in the development and subsequent transfer of the concept of good governance, aiming thereby to make international aid more effective in developing countries. New forms of governance are emerging within a particular sociopolitical context and institutional development. The region is undergoing a dual transformation process: first, a series of changes associated with the democratization of the political regime; followed by a shift to a socioeconomic model centered on market-driven development (Zurbriggen 2011).

The implementation of these transformations brought about significant changes in the role of the State, whose sphere of action diminished, while there was a gradual increase in the participation of the private sector and Non-Governmental Organizations (NGOs) in the provision of public goods and services (Zurbriggen 2011).

Governance capacity includes: (a) the existence of intangible resources that generate public value for the functioning of public organizations and ensure governability, social support, and the mobilization of key societal actors; (b) the ability to articulate and aggregate citizen demands based on the interests and ideologies involved in decision-making processes; and (c) the impact of so-called global risks (the magnitude of the consequences of those international events or situations that, unexpectedly, may negatively affect the decisions or strategies adopted by public organizations) (García 2021).

It is also necessary to develop governance capacities to manage the multiplicity of actors involved in social dynamics, and to have the ability to establish lasting agreements that facilitate the exchange of strategic resources toward achieving common goals. Governance capacities provide the relational component to the concept of institutional capacity, that is, the skills and resources available to organizations to achieve their objectives in alignment with the values assigned by society (García 2021).

Governance in public health, according to Hufty (2010), should be considered from a perspective that goes beyond technical considerations (the application of efficient procedures for a given purpose). According to Jörgensen (2005), governance includes the analysis of the rules governing the production, distribution, and consumption of the public good of health, the actors and their interactions, and allows for the identification of governmental capacities and the social support required to achieve the intended impacts of social policies.

From this perspective, governance is a strategic component for the management of the health system, but it is necessary to analyze how the different actors in the health system (providers) and civil society (users and community leaders) organize and interact to produce, distribute, and consume the public good of health according to the demand and health needs of the user population.

A turning point in the development of health governance is the basic principles promoted by the WHO since (2012): (1) coordination among all actors who can contribute to improving health; (2) collaboration among them, making health a common goal for both government and non-governmental actors; and (3) promoting the co-production of knowledge, with active participation from different actors, sharing and utilizing such knowledge. In addition, it proposes that "smart management" for health should consist of five interrelated aspects: (a) collaboration among the public, private, social, and academic sectors; (b) citizen participation; (c) a combination of government regulation and persuasion; (d) independent bodies and expert panels; and (e) adaptive policies, flexible and forward-looking structures (Kickbusch and Gleicher 2014).

7.3 Characterization of the Health Sector in Mexico

The structural characteristics of the Mexican economy, with a high level of labor informality (over 55% of employment, representing 25% of GDP [Statista 2023]), mean that public health centers are the only source of medical care for millions of people. In 2018, around 16% of the population (approximately 20 million people) lacked access to healthcare because they were not affiliated with any social security system, did not have private insurance, and were not covered by Seguro Popular (UNDP 2023).

An example of this is that, in 2016, for the first time in history, Mexico declared a health emergency due to obesity and diabetes, two non-communicable diseases that, in addition to their high cost to public finances, the epidemiological transition, and demographic change, are expected to increase health spending at a rate of 4.2% annually; furthermore, chronic-degenerative diseases will account for 90% of projected expenditures. Therefore, it is necessary to promote health prevention and promotion policies to contain the projected incidence levels of chronic diseases (Schoenfeld, 2019).

The Health System in Mexico is divided into two sections: one involving public actors (executive and legislative branches); and another, the operational part, which includes the different health systems (IMSS, ISSSTE, Ministry of Health, SEDENA, PEMEX, and private healthcare systems) and can be classified into primary, secondary, and tertiary levels (Fig. 7.1). Likewise, it changes every six years according to the sitting president, which hinders the evolution of implemented systems and, instead, leads to a complete restructuring, with the expectation of implementation, execution, and results within that period (Table 7.1).

According to Hufty's model (2010), Fig. 7.1 shows that, in the case of Mexico, the environment influences governance. Although the UN recognized health as a human right in 1946, in Mexico it was not elevated to constitutional status until 1983. Thus, every citizen and foreign national has the right to health, making migration, population growth, mobility, and economic activity phenomena that place pressure on the health system and its governance to meet the national and international goals set by the WHO.

In order to achieve these goals, they must first be established within a regulatory framework that facilitates their fulfillment. In this regard, the amendment to the General Health Law in 2019, Article 77 bis, was intended to provide the means for the entire population to have access to the right to health. However, this objective has not yet been achieved, as the data show.

This issue is exacerbated by the lack of homogeneous, complete, and verifiable information to support decision-making or interventions by the State, government agencies, or even society. Additionally, the health system and even the IMSS have

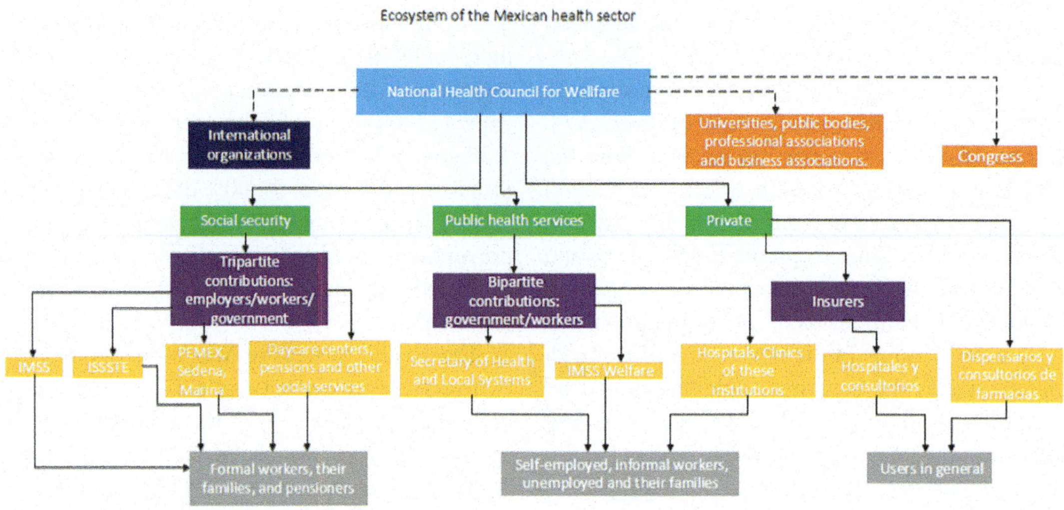

Fig. 7.1 Ecosystem of the Mexican health sector. *Source* Own elaboration

Table 7.1 Historical changes in the health sector

Reference period and name of executive	Main changes in the health sector
1990 (Carlos Salinas)	Focused on primary care and decentralization in the provision of health services
1997 (Ernesto Zedillo)	Creation of the Family Health Insurance, which allowed families to voluntarily enroll in IMSS
2003 (Vicente Fox)	Establishment of the Social Health Protection System and the creation of Seguro Popular, guaranteeing 287 essential interventions at no cost
2006–2018 (Felipe Calderón and Enrique Peña)	Implementation and reforms to the Seguro Popular System
2020–2021 (Andrés López)	The Seguro Popular was dissolved and the INSABI was established, aiming to convert the entire health system not affiliated with IMSS-ISSSTE-PEMEX-SEDENA into an IMSS-Bienestar system

Source Tovar-Pérez (2021)

their own data, which are rarely updated, and the main statistical authority in Mexico, INEGI, also lacks the statistics needed to support analysis and the generation of value-added information.

7.4 Health Indicators in Relation to the SDGs

Discussion of health care and its indicators often relies on estimates, due to the lack of reliable and timely data and control in the country. Based on SDG target 3.8, which addresses coverage of essential health services and reduction of out-of-pocket health expenditures by families, the information problem becomes evident when reviewing health coverage in Mexico as reported by INEGI (Table 7.2). According to this information, as of the 2015 review, just over 20% of the population had dual health coverage, including access to private medical services (INEGI 2020). However, starting in 2020, only one type of health coverage could be reported, assigning the burden of private medical services to Seguro Popular.

It is important to note that in 2020, Seguro Popular was discontinued and replaced by the IMSS-Bienestar service; however, the January 2024 report from the same program indicated only 11,051,574 registered users, representing just 8.42% of Mexico's 131.1 million inhabitants

Table 7.2 Evolution of population affiliation to health services

Period	IMSS (%)	Seguro Popular (%)	ISSSTE (%)	Private medical services (%)	PEMEX, SEDENA (%)	Others (%)	IMSS-Bienestar (%)	Total (%)
2010	48.8	36.2	9.9	23.98	1.5			120.38
2015	39.18	49.9	7.70	23.10	1.15	1.54		122.59
2020	51.03	35.47	8.83	2.82	1.28	1.24	1.03	101.73

Source INEGI (2020)

(IMSS 2024; CONAPO 2023a, b). It should be noted that, although the IMSS-Bienestar system has, since 2020, guaranteed that any Mexican national can request care at IMSS or IMSS-Bienestar units, individuals are not officially registered unless a medical record is created and follow-up is provided (Secretaria de Salud 2020).

On the other hand, IMSS reported having 74,045,240 beneficiaries (excluding IMSS-Bienestar) in 2023, which represents 56.47% of the total population. Therefore, if we consider the 1.28% (PEMEX, SEDENA), 8.83% (ISSSTE), 56.47% (IMSS), and 8.42% (IMSS-Bienestar), only about 75% of the population has some form of registered health coverage; this highlights the lack of universal health service coverage in the country (INEGI 2020; IMSS 2023).

In contrast to the data from INEGI and IMSS, the 2021 National Health and Nutrition Survey (ENSANUT) shows that only 23.2% of the population seeks care at IMSS, 16.4% at other government health services (SEDENA, PEMEX, Ministry of Health services), and 5.5% at private hospitals (Rodríguez 2022). This means that 45.1% of care is provided in reportable services, while the remaining information on care, diagnoses, medications used, and more is generated in private medical services—whether in offices, home consultations, or pharmacy-adjacent clinics—which do not report their data and, therefore, this information is lost without being recorded.

Based on the above, it can be stated that among reportable services, there are 41,783 clinics available nationwide, including medical, psychological, dental, and other services, with 70.5% located in primary care settings. However, only 51.6% of all clinics (21,564) are first contact (general and family medicine), representing 6,033.2 inhabitants per clinic (Secretaria de Economía 2023). Assuming an average of 40 patients per clinic per day, it would take 151 days, or about 30 weeks, to provide a single consultation to the entire population, without follow-up.

As of August 2022, at least 340,000 daily consultations were identified in clinics attached to registered pharmacies (Rodríguez 2022). In terms of expenditures, pharmacy sales in 2017 totaled 4.132 billion dollars; meanwhile, average household spending on medications reached 1,345 pesos in 2023, representing a 31% increase compared to 2018, according to the National Household Income and Expenditure Survey (Tejeda 2018; INEGI 2023).

For the review of other health-related SDG indicators, reports from the Epidemiological Bulletin of the National Epidemiological Surveillance System (SINAVE) of the General Directorate of Epidemiology (DGE),[1] the World Health Organization (WHO), and the SDG reporting system of the Government of Mexico were used as references (DGE 2024a, b; DGE 2024b; INEGI 2022; DGE 2015; CONAPO 2023a; WHO 2021; Government of Mexico 2023), using 2015—the year the SDGs were signed—as the baseline for comparison (Table 7.3).

After reviewing the data related to the SDGs, it is important, first, to note that the objectives *3.5.1 Coverage of treatment for substance use disorders* and *3.9.1 Mortality rate attributable to pollution* are not reported in the statistical or SDG

[1] Department of the Ministry of Health responsible for providing epidemiological information and knowledge that supports public policies in the sector.

Table 7.3 Comparison of SDG health indicators: 2015 versus present

Indicators	2015	Most recent data
3.1.1 Maternal mortality	810	556°
3.1.2 Births attended by skilled health personnel	94.63	85.1 *
3.2.1 Under-five mortality rate	16.01	14.78 *
3.2.2 Neonatal mortality rate	8.72	7.93°
3.3.1 Number of new HIV infections	19,103	20,342°
3.3.2 Tuberculosis incidence per 100,000 population	13.55	12.4 *
3.3.3 Malaria incidence per 1,000	0.42	0.19 *
3.3.4 Hepatitis B incidence per 100,000	747	837°
3.3.5 Number of people requiring interventions against tropical diseases	517	246 *
3.4.1 Mortality rate attributed to cardiovascular diseases, cancer, diabetes, or chronic respiratory diseases	308.05	386.63 *
3.4.2 Suicide mortality rate °	6,425	8,239°
3.5.2 Per capita alcohol consumption	5.2	5.7 (&)
3.6.1 Mortality rate due to road traffic injuries	13.6	12.8 (&)
3.7.1 Percentage of women of reproductive age using modern family planning methods	81.55%	83.14% *
3.7.2 Adolescent pregnancy rate per 1,000 adolescents	72.4	60.3°
3.9.2 Mortality rate attributable to unsafe water, inadequate sanitation, and lack of hygiene		855 *
3.9.3 Mortality rate attributable to unintentional poisonings		0.36 *

Notes: °2022 data; *2021 data; (&) 2020 data
Source DGE (2024a, b),INEGI (2022), DGE (2015), CONAPO (2023a), WHO (2021), Government of Mexico (2023)

reporting systems; and, second, that the mortality rate attributable to unsafe water and voluntary poisoning does not have an initial comparison point since reporting begins with 2021 data.

Furthermore, as previously noted, there is no reliable data regarding the SDG indicators. For example, in the case of maternal mortality, according to the Government of Mexico's 2030 Agenda report, there were 1,109 maternal deaths nationwide in 2021 compared to 1,036 in 2021 according to the DGE (DGE 2022; Government of Mexico 2023). It is not possible to assert that the information is fully transparent and complete, which means a key element in health governance—information—is lacking, and consequently, managerial capacity for resource allocation is inefficient.

Regarding progress on the SDGs and the targets for maternal mortality, women's health care, and neonatal mortality, the figures have decreased significantly, in line with global reports. However, when reviewing data such as new HIV infections, alcohol consumption, suicide, cardiovascular diseases, cancer, diabetes, or chronic respiratory diseases, these have increased in the country, which is alarming. Moreover, progress is contrary to the global situation regarding tuberculosis, according to WHO reports.

Thus, it is evident that efforts in Mexico have focused on maternal health and infectious diseases such as malaria, tuberculosis, and tropical diseases. Based on this, it can be generally stated that (a) the information available in Mexico is incomplete and unreliable; (b) except for objectives 3.1 and 3.2, the remaining health-related SDG targets will not be met and, on the contrary, there is regression; (c) there is no evidence of adequate management capacity regarding available resources and the generation of material resources to improve care; and (d) it is necessary to identify how decision-making is carried out regarding SDG-related expenditures.

Table 7.6 Percentage of budget investment in the SDGs by agency

Agency	Percentage allocated to the SDGs (%)
SEDENA	99.9
SSA	94.7
SEMAR	100
IMSS	79.5
ISSSTE	75.3
PEMEX	99.4

Source CIEP (2022)

It should be noted that the Government of Mexico-INEGI's 2030 Agenda reports do not include or consider the vaccination indicator. However, with the data generated from the versions of ENSANUT, it can be noted that in 2021, 31.1% of children under two years of age had a complete vaccination schedule, a percentage that dropped to 26.6% in 2022. This situation is particularly alarming considering that in 2012 the percentage was 74.2% and in 2006 it was 78.4% (Mongua-Rodriguez 2023; Codigo 2022).

Regarding progress toward the SDGs at the global level compared to the results in Mexico, the 2023 report addresses the advances in these objectives (yellow indicates negative, green indicates positive, Table 7.4).

After reviewing health indicators related to the SDGs, an analysis was conducted to determine whether there was alignment between government investment and targets to achieve these objectives. Table 7.5 shows the historical investment based on the percentage of Gross Domestic Product (GDP) allocated to health (Statista 2023; Banxico 2023).

Additionally, according to reports from the Center for Economic and Budgetary Research (CIEP 2022), except for SEDENA, SEMAR, and PEMEX, their budgets are aligned with the SDGs for 2021, as shown in Table 7.6. It is important to note that these institutions have the lowest percentage of national coverage. Furthermore, in the case of IMSS and ISSSTE, these institutions face their own specific challenges.

On the other hand, Fig. 7.2 shows the percentage of health spending allocated to achieving any target of SDG 3, revealing that, in particular, target 3.c (which aims to improve health financing and increase the number of qualified health personnel for care) is especially low, at only 1%.

7.5 The Challenges of the SDGs in Mexico

The importance of addressing governance capacities in the health sector lies in the review of components and indicators aimed at analyzing the system's steering capacity from the perspective of its political actors, a dynamic that is scarcely described in the public health literature in Mexico. The perspective of political actors stems from analytical approaches that highlight the need to address and understand their agendas, a skill that must be complemented by the ability to both generate and utilize strategic information for health policy management (Arredondo et al. 2005).

In this context, the main challenges facing the Mexican health system are to ensure access to health services and to reduce household spending on them. These challenges are the result of the historical convergence of various economic, social, and political factors. Reforms have sought to address needs and social demands, but the results have not been consolidated due to diverse and complex circumstances.

The success of a public policy in health promotion, disease prevention, and timely diagnosis will only be possible through the joint participation of both citizens and government. The multidimensional and multicausal nature of the complex health issues in Mexico requires an integrated response that makes health a shared goal of both government and society (Schoenfeld, 2019).

In the many efforts in the field of health promotion and the prevention of obesity and non-communicable chronic diseases in Mexico, mainly driven by the federal and local governments, organized civil society also participates through health research centers, foundations, the business sector, and other stakeholders. Despite their reach, these efforts and programs remain isolated and fragmented, preventing coordinated

Table 7.4 Comparison of health-related SDG results in Mexico versus global results

Indicator	2015	Most recent data	Mexico	Global average
3.1.1 Maternal mortality	810	556°	Decrease	Decrease
3.1.2 Births attended by skilled health personnel	94.63	85.1 *	Decrease	Increase (including in Latin America)
3.2.1 Under-five mortality rate	16.01	14.78 *	Decrease	Decrease
3.2.2 Neonatal mortality rate	8.72	7.93°	Decrease	Decrease
3.3.1 Number of new HIV infections	19,103	20,342°	Increase	Decrease
3.3.2 Tuberculosis incidence per 100,000 population	13.55	12.4 *	Decrease (−8.49%)	Decrease (−10.67%)
3.3.3 Malaria incidence per 1,000	0.42	0.19 *	Decrease	Decrease
Vaccination			Decrease	Decrease
Health coverage			Decrease	Increase 3%

Source Own elaboration and WHO data (2021) Notes: ° 2022 data; * 2021 data

Table 7.5 Historical GDP investment in health in Mexico

Year	% GDP	Per capita (USD)	Change from previous year (%)	Annual inflation (%)
2022	2.91	6,710	+3.26	7.82
2021	3.05	6,498	+2.61	7.36
2020	3.29	6,333	+7.25	3.15
2019	2.68	5,905	−1.94	2.83
2018	2.67	6,022	−1.97	4.83
2017	2.76	6,143	−1.29	6.77
2016	2.85	6,223	−3.65	3.36
2015	2.99	6,459		2.13

Source Statista data (2023) and Banxico (2023)

action and, in some cases, resulting in duplication of efforts and increased costs due to a lack of communication among stakeholders. As a result, these initiatives have not had an effective impact on health promotion or disease prevention (Schoenfeld, 2019).

Mexico established a high-level body called the National Council for the 2030 Agenda, led by the Office of the President and bringing together multiple sectors of the country: academia, the private sector, civil society, among others. It also has a Specialized Technical Committee on the SDGs, focused on developing conceptual, theoretical, and methodological tools for generating information to facilitate monitoring of the 2030 Agenda implementation strategy, and it includes various coordination mechanisms between the federal government and local governments.

Although there is a specific council to generate information on the targets set in the SDGs, according to the WHO platform, Mexico's information is not up to date. There is also a disparity between the budget and the number of users to be served. Thus, the budgets of IMSS and ISSSTE, relative to the number of users they must serve, are not comparable to the social security systems of PEMEX and SEDENA.

Changes in public policy with each presidential term reflect the core problem of a lack of integration of efforts and collaborative approaches among stakeholders, which would enable a better health system that meets the targets set in the

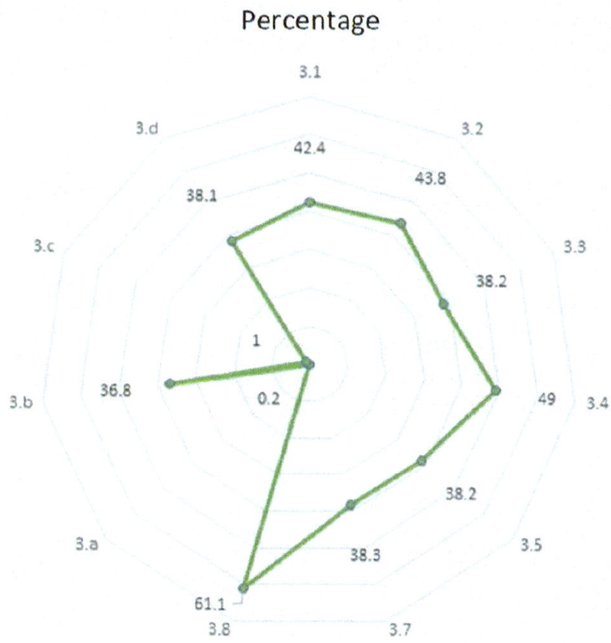

Fig. 7.2 Radar chart of the percentage of spending allocated to achieving SDG 3 targets. *Source* CIEP (2022)

SDGs and that Mexico must achieve to realize universal access to health.

The lack of governance and existing external factors have resulted in a lack of access to health care for the entire population, especially for those who are not entitled to social security, and even among those who are, there is a high out-of-pocket expenditure on private health care.

7.6 Conclusions

Governance capacity is a key variable in the health system that enables the identification of essential dimensions (actors, processes, and regulations) aimed at ensuring the highest standards in the design, prioritization, implementation, and monitoring of actions, programs, and policies that respond to the health needs of the population (Arredondo et al. 2021).

Governments cannot manage the magnitude of global risks in the implementation of the 2030 Agenda and achievement of the SDGs in isolation or unilaterally. To do so, they must develop new models of governance that support risk management tasks by establishing networks and institutional mechanisms that facilitate the exchange of knowledge and strategic resources (García 2021).

If necessity is the mother of invention, then the same forces that create current uncertainties may also provide the means to address them. Uncertainty creates the possibility for change, even for the better.

Within the governance framework of the Mexican health sector, the Ministry of Health, as the lead agency, must establish levels of agreement and coordination with all actors involved in the organized social response. To achieve better levels of governance, mechanisms are needed to promote interaction, participation, and dialogue among public and social actors and institutions.

Among the recommendations proposed in the reviewed literature are:

a. Establish operational rules and regulations to increase the feasibility of any health program with the support of all stakeholders involved.
b. Promote the establishment of agreements among all actors in the ecosystem to guide strategies that address the objectives of the health plan.

c. Create a citizen and health observatory to monitor indicators and governance processes in health.
d. Regardless of the health plan, it is necessary to develop and validate mechanisms that promote and guarantee the participation of all stakeholders in decision-making regarding national, state, and local health issues.
e. Formalize state and local consultation forums with all the actors for the design, monitoring, and evaluation of health programs, particularly in relation to achieving the SDGs.
f. Create binding spaces for accountability at the national, state, and local levels, providing accessible information on health indicator outcomes and the efficient use of human and financial resources by disease type.

Acknowledgements This research was funded by the Instituto Politécnico Nacional through the Secretaria de Investigación y Posgrado, projects SIP 20240589 and SIP20241937.

References

Arredondo A, Orozco E, Castañeda X (2005) Gobernanza en sistemas de salud y participación social en México a partir de la reforma del sector. Hitos de Ciencias Económico Administrativas 11(29):5–12

Arredondo A, Orozco E, Cuadra M, Torres Ch, Reséndiz O, Castrejón B (2021) Governance in the Health System, INSP. https://insp.mx/assets/documents/webinars/2021/CISS-Gobernanza.pdf

Banxico (2023) Portal de inflación. Retrieved December 1, 2023, from https://www.banxico.org.mx/tipcamb/main.do?page=inf&idioma=sp

CIEP (2022) Health expenditure and sustainable development goals. https://ciep.mx/wp-content/uploads/2022/10/Gasto-en-Salud-y-Objetivos-de-Desarrollo-Sostenible.pdf

Codigo F (n.d.) Full vaccination coverage in two-year-old children drops: from 31.1% in 2021 to 26.6% in 2022. Código F. https://codigof.mx/cae-cobertura-de-vacunacion-completa-en-ninos-de-dos-anos-de-31-1-en-2021-a-26-6-en-2022/

CONAPO (2023) World population day. Population Projections for Mexico for the Next 50 Years: 2020–2070. Government of Mexico. https://www.gob.mx/conapo/prensa/dia-mundial-de-la-poblacion-las-proyecciones-de-la-poblacion-de-mexico-para-los-proximos-50-anos-2020-2070?idiom=es#:~:text=En%20el%20marco%20del%20D%C3%ADa,del%20Consejo%20Nacional%20de%20Poblaci%C3%B3n.&text=El%20volumen%20de%20la%20poblaci%C3%B3n,es%20de%200.9%20por%20ciento.

CONAPO (2023a) Adolescent fertility rate drops by more than 16 percent. CONAPO. https://www.gob.mx/conapo/prensa/deciende-mas-del-16-por-ciento-la-tasa-de-fecundidad-de-adolescentes-septiembre-2023?idiom=es#:~:text=de%20septiembre%202023-,En%20el%20marco%20del%2026%20de%20septiembre%2C%20D%C3%ADa%20Mundial%20de,de%2015%20a%2019%20a%C3%B1os).

Consejo Nacional de Evaluación de la Política de Desarrollo Social (Coneval) (2023) Pobreza multidimensional 2022. CONEVAL. https://www.coneval.org.mx/Medicion/MP/Documents/MMP_2022/Pobreza_multidimensional_2022.pdf

DGE (2015) Semana epidemiológica 52 (2015) Histórico Boletín Epidemiológico [Conjunto de datos]. En Dirección General de Epidemiología. https://www.gob.mx/salud/acciones-y-programas/historico-boletinepidemiologico

DGE (2022) Semana epidemiológica 52 (2021) Histórico Boletín Epidemiológico [Conjunto de datos]. En Dirección General de Epidemiología. https://www.gob.mx/salud/acciones-y-programas/historico-boletinepidemiologico

DGE (2024a) Historical epidemiological bulletin. In: General directorate of epidemiology. https://www.gob.mx/salud/acciones-y-programas/historico-boletin-epidemiologico

DGE (2024b) Weekly reports for the epidemiological surveillance of maternal deaths 2023. In: General directorate of epidemiology. https://www.gob.mx/salud/documentos/informes-semanales-para-la-vigilancia-epidemiologica-de-muertes-maternas-2023

Díaz-Castro L, Ramírez-Rojas M, Cabello-Rangel H, Arrendondo A, Báscolo E (2021) Governance in the health system in response to the COVID-19 Pandemic in Mexico. Salud Pública Mex 63(5):672–681

García R (2021) Global risks and governance capacities. Keys to the implementation of the 2030 agenda. Review of CLAD Reforma y Democracia, No. 79, pp 39–76, ISSN: 1315-2378

Government of Mexico (2023) Sustainable development goals. In: Agenda 2030. https://agenda2030.mx/#/home

Hufty M (2010) Gobernanza en salud pública: hacia un marco analítico. Revista de Salud pública 12:39–61

IMSS (2023) Report to the federal executive and the congress of the union on the financial situation and risks of the Mexican social security institute (1st ed.). https://www.imss.gob.mx/sites/all/statics/pdf/informes/20222023/19-informe-completo.pdf

IMSS (2024) Registered population roster 2024 IMSS-Bienestar Program. In: IMSS Bienestar (2024 Edition). https://www.imss.gob.mx/sites/all/statics/imssBienestar/estadisticas/01-PoblacionAtendida-2024.pdf

INEGI (2022) Mental Health. In: INEGI. https://inegi.org.mx/temas/salud/

INEGI (2023) National Household Income and Expenditure Survey 2022 (ENIGH). In: ENIGH. https://www.inegi.org.mx/contenidos/programas/enigh/nc/2022/doc/enigh2022_ns_presentacion_resultados.pdf

INEGI (n.d.) Social security coverage. In: INEGI (2020 Edition)

Jørgensen TB, Vrangbæk K (2005) Det offentlige styringsunivers: Fra government til governance? (1st ed.) Magtudredningen. https://unipress.dk/media/14525/8779348807_det_offentlige_styringsunivers.pdf

Kickbusch I, Gleicher D (2014) Smart governance for health and well-being: the evidence. World Health Organization

Ministry of Economy (2023) Physicians, nurses, and other health specialists. In: DATA MEXICO (2023-Q2 Edition). https://www.economia.gob.mx/datamexico/es/profile/occupation/medicos-enfermeras-y-otros-especialistas-en-salud

Ministry of Health (2020) 002. Care for people without social security is guaranteed in public health Centers: Insabi. Government of Mexico. https://www.gob.mx/salud/prensa/002-atencion-a-personas-sin-seguridad-social-se-garantiza-en-los-centros-publicos-de-salud-insabi

Mongua-Rodríguez N, Delgado-Sánchez G, Ferreira-Guerrero E, Ferreyra-Reyes L, Martínez-Hernández M, Canizales-Quintero S, … García-García L (2023) Vaccination coverage in children and adolescents in Mexico. *Salud Pública de México* 65:1–11

Rodríguez A (2022) Pharmacy medical offices in Mexico: how many people do they serve per day? *El Financiero*. https://www.elfinanciero.com.mx/empresas/2022/08/18/consultorios-medicos-de-farmacias-en-mexico-cuantas-personas-atienden-al-dia/

Santos L (2022) Human rights and sustainable development goals in the global governance of new technologies for human health. Medicina y Ética 33(4):1039–1060

Secretaría de Economía (2023) Médicos, enfermeras y otros especialistas en salud [Perfil de ocupación]. DataMéxico. https://www.economia.gob.mx/datamexico/es/profile/occupation/medicos-enfermeras-yotros especialistas-en-salud

Secretaría de Salud (2020, 3 enero) Atención a personas sin seguridad social se garantiza en los centros públicos de salud: INSABI. Gobierno de México. https://www.gob.mx/salud/prensa/002-atencion-a-personassin-seguridad-social-se-garantiza-en-los-centros-publicos-de-salud-insa

Schady N, Holla A, Sabarwal S, Silva J, Yi A (2023) Collapse and recovery: how the Covid-19 Pandemic eroded human capital and what to do about it. World Bank Group

Schoenfeld EN (2019) Hacia un nuevo modelo de gobernanza para la promoción de la salud. Buen Gobierno (26):1–28

Statista (n.d.) Mexico: health expenditure per capita 2010–2022. https://es.statista.com/estadisticas/1150470/gasto-en-salud-per-capita-porcentaje-pib-mexico/

Tejeda C (2018) Farmacias Similares and Guadalajara, the "Healthiest" in the Market. Fortuna y Poder. https://fortunaypoder.com/corporativos/farmacias-similares-guadalajara-las-mas-sanas-del-mercado

Tovar Pérez GD (2021) Evolución y actualidad del sistema de salud de México: Una comparación con el sistema sanitario cubano. Epikeia. Revista del Departamento de Ciencias Sociales y Humanidades (42):1–24

UNDP (2022) Human development report 2021/2022, United Nations Development Programme

UNDP (2023) Mapping the socioeconomic consequences of COVID-19 in Latin America and the Caribbean and the Responses Adopted for Recovery, UNDP

UNESCO (2015) Concept of governance, International Bureau of Education, http://www.ibe.unesco.org/en/geqaf/technical-notes/concept-governance

WHO (2021) Expert advisory committee on developing global standards for governance and oversight of human genome editing. Human genome editing: a framework for governance. https://who.int/publications/i/item/9789240030060

WHOdata (2024) WHO COVID-19 Dashboard, https://data.who.int/dashboards/covid19/deaths?n=c. Retrieved 01/20/2024

World Health Organization (2012) Governance for Health in the 21st Century. Ed. The Regional Office for Europe, Geneva, Switzerland

Zurbriggen C, Governance: a perspective from Latin America. Perfiles Latinoamericanos 19(38)

A Systemic Approach to Social Entrepreneurship and Its Relationship with Sustainable Water Management in Tourism Dynamics

8

Zeltzin Pérez-Matamoros, Ricardo Tejeida-Padilla, Roberto Patiño-Abuela, and Isaías Badillo-Piña

Abstract

In response to the challenges of tourism in environments under high pressure on natural resources, social entrepreneurship emerges as a strategy capable of generating social innovation while contributing to water and territorial sustainability. Water management and conservation are vital aspects of life, underscoring their importance for ecosystem health and the basic needs of society. The Sustainable Development Goals (SDGs) guide tourism practices towards continuous improvement. The aim of this research is to identify and interpret the interaction between water, social entrepreneurship and tourism dynamics in rural communities of Mexico City, providing a comprehensive diagnosis of the current situation. To achieve this, the Total Systems Intervention (TSI) was applied. The findings reveal that social entrepreneurship has a positive impact on sustainable development and the preservation of water resources, being essential for long-term sustainability in tourism dynamics.

Z. Pérez-Matamoros · I. Badillo-Piña
SEPI ESIME-Zacatenco, National Polytechnic Institute, Mexico City, Mexico

R. Tejeida-Padilla (✉) · R. Patiño-Abuela
SEPI EST, National Polytechnic Institute, Mexico City, Mexico
e-mail: rtejeidap@ipn.mx

Keywords

Water · Sustainability · Social entrepreneurship · Systems thinking

8.1 Introduction

Sustainable development presents a fundamental challenge for the twenty-first century, with both direct and indirect implications in social, economic and environmental spheres (Zamanzad-Ghavidel et al. 2021). Rapid urbanization and climate change have led to significant changes in hydrological patterns and the frequency of extreme weather events (Boroomandnia et al. 2021). These changes have altered the natural peaks and volumes of surface runoff, which in turn cause degradation in water quality, environmental systems and ecosystems (Babaei et al. 2018).

In this context, water management is essential for sustainable development, as water is a vital resource for all forms of life and its scarcity can have significant consequences for the environment, society and the economy. Proper management is crucial to ensure sustainability and overall well-being. Various studies have highlighted the importance of water management in key sectors such as agriculture, industry and tourism (Chopra and Ramachandran 2021; Dwyer 2022; Nhemachena et al. 2020).

P. Pérez-Hernandez and R. Huerta Cuervo (eds.), *Social Well-Being and the Challenges in the Sustainable Development Goals*, Sustainable Development Goals Series,
https://doi.org/10.1007/978-3-032-14294-8_7

The constant evolution and growth of tourism has become a key pillar for progress towards the Sustainable Development Goals (SDGs) established in the 2030 Agenda, as they represent a universal call to action to end poverty, protect the planet and improve lives worldwide (UN 2024a). Thus, tourism is presented as a significant tool for advancing the SDGs by promoting responsible practices that manage water use in tourist destinations, fostering environmental awareness and education among tourists, and contributing to the economic development of local communities in an inclusive and equitable manner (Wu and Liang 2023).

The adoption of policies and measures aimed at sustainability in the tourism sector not only benefits present generations but also lays the foundation for a more prosperous and balanced future for generations to come. The current context of tourism dynamics highlights the need to adopt an integral and systemic approach to address challenges related to water management (Bouras et al. 2020; Vuković 2022).

This research aims to study and interpret social entrepreneurship from a systemic perspective, in order to identify its potential to generate innovative solutions to social and environmental issues, particularly in its connection with sustainable water management in tourism contexts. To this end, the Total Systems Intervention (TSI) (Jackson 2019) is used as a conceptual tool to understand the complexity of these dynamics and their impact on local communities and their environment.

1. Social Entrepreneurship in Tourism

In recent years, social entrepreneurship has emerged as a dynamic and adaptable approach to address the various challenges faced by the tourism sector in terms of sustainability and social responsibility. It is distinguished by its capacity to innovate in the resolution of long-standing socio-environmental problems (Khan et al. 2021; Kummitha et al. 2021). This characteristic makes it a powerful tool for promoting sustainable water management in tourist destinations, by offering creative solutions that engage local communities in participatory decision-making processes for the conservation and efficient use of water.

The convergence of social entrepreneurship and tourism creates a favourable environment for the development of innovative solutions that go beyond conventional approaches in the tourism sector. This intersection opens new opportunities for the creation of inclusive and sustainable business models that benefit both tourists and host communities (Sharma and Bhat 2023; Tleuberdinova et al. 2021). Likewise, social entrepreneurship in tourism can serve as a catalyst for job creation and the improvement of residents' quality of life, while also contributing to the conservation of natural resources (Gutierrez and Vafadari 2023).

From this perspective, social entrepreneurship initiatives can become spaces for learning and connection with the environment, fostering greater environmental awareness among visitors to these destinations (Li et al. 2022; Papaluca et al. 2020). Through tourism experiences that promote responsibility and environmental education, travellers can gain knowledge about the importance of water conservation in tourist destinations and the measures they can take to reduce their water footprint during their trips (Aman et al. 2021; Paudyal 2019). This can not only help foster a culture of responsible tourism but also inspire individual and collective actions for the protection of water resources globally. From a systemic perspective, social entrepreneurship in tourism contributes to multiple levels of action that directly impact water sustainability, as shown in Fig. 8.1.

The integration of social and environmental responsibility principles into social entrepreneurship within the tourism sector has the potential to foster equity and participation, reduce socioeconomic gaps and strengthen the adaptive capacity of communities in the face of climate change and other current environmental challenges (Rela et al. 2020).

In many tourist regions, local communities face challenges related to limited access to safe drinking water and basic sanitation. Social enterprises can work in collaboration with these communities to develop innovative solutions that improve access to safe water and contribute to sustainable economic and social development

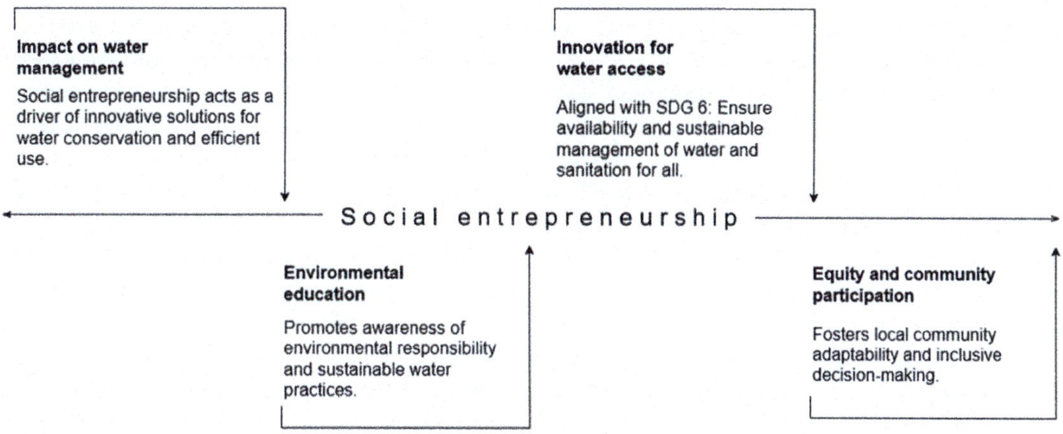

Fig. 8.1 Interconnected pathways linking social entrepreneurship and sustainable water use. *Source* Own elaboration

Table 8.1 SDGs related to social entrepreneurship and water management

SDG number	Purpose	Relationship with water
6. Clean water and sanitation	Improving the efficient use of water resources as a key to reducing water stress	Implementation of integrated water resources management at all levels
8. Decent work and economic growth	Improve employment opportunities, especially among youth, reduce informal employment and labour market inequality, promote safe and secure working environments, and enhance access to financial services to ensure sustained and inclusive economic growth	Inclusive and sustained economic growth to drive progress and improve living standards
13. Climate action	Immediate measures to prevent catastrophic environmental consequences and ensure a sustainable future for coming generations	Education, awareness and human and institutional capacity regarding climate change mitigation, adaptation, impact reduction and early warning

Source Own elaboration

(Cockburn-Wootten and McIntosh 2020; Picciotto and Ummarino 2022).

2. Water and Social Entrepreneurship in the SDGs

Water, essential source of life, is one of the global priorities in the 2030 Agenda, addressed in SDG 6: Clean Water and Sanitation. The management of this water resource is fundamental, as it has a direct impact on the viability of entrepreneurial initiatives. In this context, due to the virtues inherent in social entrepreneurship, it is also possible to link it with other SDGs, such as SDG 8: Decent Work and Economic Growth (UN 2024c, 2024b).

Initiatives arising from this approach seek to address social and environmental issues in an integrated manner, establishing direct links with sustainable water management as part of a broader commitment to sustainable development. For a better understanding of how some SDGs are involved with the subject of study, Table 8.1 is presented below.

Considering the SDGs as a reference framework can strengthen strategies aimed at more responsible water management, especially when linked to the involvement of social enterprises. These initiatives can play a strategic role in promoting sustainable solutions tailored to the realities of each region. In addition to addressing threats to ecosystems, these enterprises contribute to achieving SDGs 6, 8 and 13. In this way, the long-term sustainability of water resources and human well-being is ensured.

Additionally, innovation and technology play a key role in the development of social enterprises by facilitating the creation and adoption of scalable and accessible solutions capable of reaching marginalized and vulnerable communities (Skivko 2021; Xiao and Su 2022). From affordable water purification systems to mobile applications for water quality monitoring, technology is being creatively used to address challenges related to access, quality and sustainable water management in various contexts (Popkova et al. 2023).

Multi-stakeholder collaboration and partnership building are fundamental aspects for the success of social enterprises related to water management. The active participation of governments, businesses, non-governmental organizations and local communities is essential to identify needs, mobilize resources and create an enabling environment for innovation and collective action. Moreover, promoting inclusive and sustainable business models that involve both public and private sector actors can contribute to building resilient and equitable water management systems.

3. The Dynamics of Tourism and Water in Rural Areas of Mexico City

The tourism system in Mexico City has had limited involvement in water resource management, as is the case in rural areas with tourism potential where communities depend on this resource for their livelihoods. Rural tourism, defined as tourism activity that takes place in rural areas and often unfolds in natural and cultural settings, can have significant effects on the availability, management and quality of water (Epuran et al. 2020; Romão 2018).

Given this scenario, it is essential to reflect on the sustainability of water resources, understood as the capacity to use water in sufficient quantity and quality to meet the needs of local communities and ecosystems, both now and in the future, with the aim of conserving and protecting the hydrological balance of watersheds (Cervantes-Jiménez et al. 2020). It is emphasized that social sustainability can have three general orientations: towards behavioural change to achieve environmental goals; towards the preservation of sociocultural patterns and practices; and towards the mitigation of poverty and inequality (Santos and Moreira 2021).

The increase in tourism places additional pressure on local water resources, especially in regions where water infrastructure is limited or inadequate. The demand for water for tourism purposes, including the supply of hotels, restaurants and recreational activities, can overload existing water supply systems and contribute to water scarcity and pollution. In addition, it generates solid and liquid waste that affects water quality and requires effective wastewater management and treatment systems (Grison et al. 2023; Liu et al. 2021).

For these reasons, there is an urgent need for comprehensive water planning and management to address the challenges associated with its use in rural areas of Mexico City. This requires adopting integrated approaches that consider both tourism needs and the objectives of conservation and sustainable water resource management. The implementation of appropriate policies and regulations, as well as the active participation of local communities and relevant stakeholders, is necessary to ensure tourism development that is compatible with the protection and preservation of water and the environment.

Mexico City, covering an area of 1,494.3 km2, is part of the Neovolcanic Axis province, representing 0.1% of the country's surface area. Divided into 16 territorial districts (Fig. 8.2), most of its territory has a temperate sub-humid climate (87%), while the rest is dry and semi-dry (7%) and temperate humid (6%). Rainfall occurs mainly in summer, with total annual precipitation varying in the dry region it is 600 mm, while in the temperate humid area it reaches 1,200 mm per year (INEGI 2024).

The Government of Mexico City and the Secretariat of the Environment (SEDEMA) created the Special Green Infrastructure Programme of Mexico City (SGIP-Mexico City). This is a territorial management tool that generates strategies and guidelines in an organized manner to achieve

Fig. 8.2 Map of Mexico City by territorial district. *Source* Own elaboration

the greatest environmental and social benefit (SEDEMA 2024).

The SGIP-Mexico City is composed of nine Green Infrastructure Regions: 1. Northern Urban Green Areas Region (NUGAR), 2. Urban Ravines Region (URR), 3. Central-Western Urban Green Areas Region (CWUGAR), 4. Central-Eastern Urban Green Areas Region (CEUGAR), 5. Wetlands Region (WR), 6. Eastern Urban Green Areas Region (EUGAR), 7. Rural–Urban Transition Region (RUTR), 8. Natural Forests Region (NFR), 9. Rocky Outcrops Region (ROP); as shown in Fig. 8.3.

The RUTR and NFR correspond to the boroughs of Álvaro Obregón, Cuajimalpa, Magdalena Contreras, Milpa Alta, Tláhuac, Tlalpan and Xochimilco (SEDEMA 2019). The representation of these regions for Mexico City highlights the importance of sustainably planning and managing the resources and ecosystems of rural communities.

Likewise, based on data from the National Statistical Directory of Economic Units, in the aforementioned boroughs, there are 8,556 registered economic establishments with a size of 0–5 and 6–10 people that are related to the dynamics of the tourism system. These economic establishments are segmented as follows (Table 8.2).

These data reflect the economic importance of tourism in the aforementioned boroughs, as well as the presence of micro-establishments that, due to their scale and proximity to the territory, have the potential to implement sustainable practices at the local level. In this context, it is relevant to study these dynamics from a systemic perspective that allows for understanding their inter-

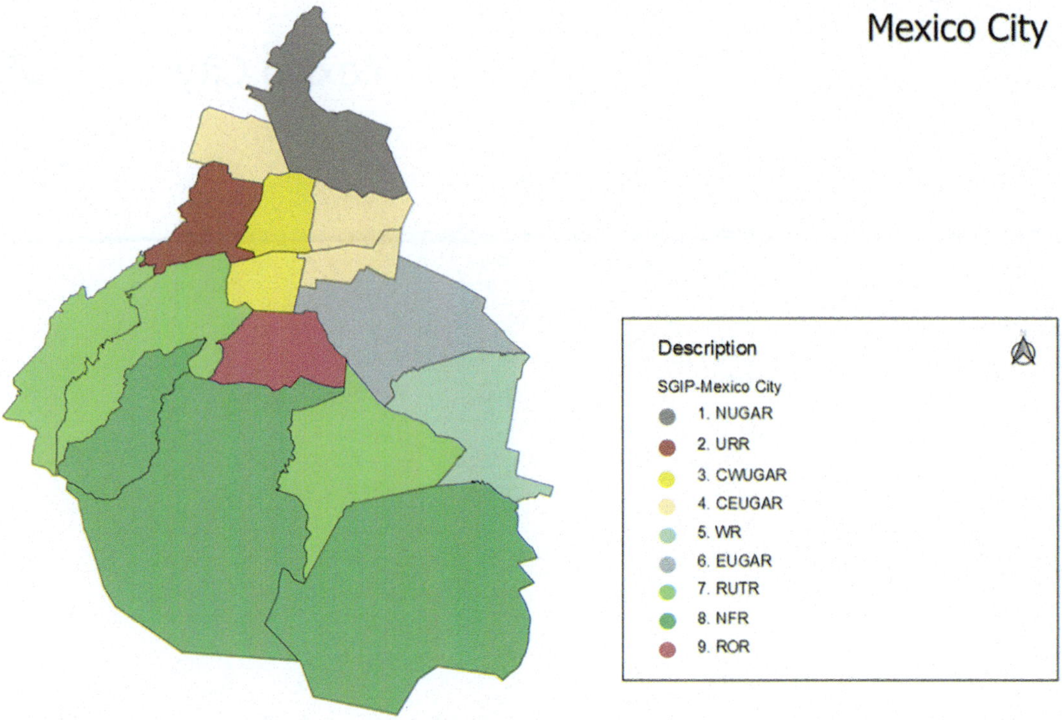

Fig. 8.3 Green Infrastructure Regions. *Source* Own elaboration

Table 8.2 Registered establishments linked to tourism activities

Economic activity	Description	Quantity
Temporary accommodation and food and beverage preparation services	• Cabins, villas and similar • Campsites and recreational hostels • Food preparation services	8,368
Other services, except government activities	• Other personal services • Regulatory associations of recreational activities	188
	Total	8,556

Source DENUE (2024)

relationships and guiding more comprehensive strategies.

8.2 Methodology

To address this complex issue and meet the research objective, the Systemic Paradigm was adopted, using the TSI. This approach enables the identification of participants' roles in decision-making processes, as well as the type of system in which the problem arises (Jackson 2019). The first step in applying TSI is to determine whether the system in question is simple or complex. Simple systems are typically static and closed, involving a small number of components and linear relationships. In contrast, complex systems consist of a large number of interconnected elements and exhibit non-linear relationships. These systems maintain a constant exchange of information with their environment and must adapt to it in order to evolve.

Subsequently, the type of relationships generated by the system's actors is identified: unitary, pluralist, or coercive. Unitary when all those involved agree with the system's teleology, their beliefs and values are compatible, and they share common interests. Decision-making is carried out with the participation of all. Pluralist, when actors have different interests, values and beliefs; however, consensus is reached to achieve the system's teleology. Coercive when there are conflictual relationships, there is little common interest among actors, and consensus is generated through the domination of one group over another.

The Context-Problem Matrix (Table 8.3) categorizes in its quadrants the type of system being addressed and the archetype of relationships that occur among the stakeholders.

As part of the methodological procedure, field visits were conducted in rural communities of Mexico City with the aim of identifying and interpreting the relationships among the involved stakeholders. During these visits, both direct and participatory observation were carried out in enterprises with identifiable links to the local tourism dynamic. The experience provided to tourists or visitors by these enterprises was also documented, as well as their relationship with entrepreneurs and the attractions involved in the dynamic, with particular emphasis on water resources.

8.3 Results

In the study and interpretation of the system, through direct and experiential observation in various enterprises related to tourism dynamics, several key components that make up the system's structure were identified, as presented in Table 8.4.

Among the identified actors and entities, there are a large number of relationships. As a human activity system, these interactions are neither linear nor unidirectional, reflecting the complex and dynamic nature of the system under study (Fig. 8.4).

In analysing the relationships generated with the actors involved in the system, social enterprises in rural communities are identified, where entrepreneurial individuals contribute their knowledge of the community's unique attractions. Social entrepreneurship is primarily characterized by its aim to address socio-environmental issues, seeking to generate a positive impact on society and the environment. In addition, these initiatives pursue innovative and sustainable solutions to challenges such as poverty, lack of access to education, health and environmental protection, specifically including water conservation in the context of tourism activities involving visitors who come to experience and enjoy the local attractions.

In this context, the following rural communities from the aforementioned boroughs were identified (Table 8.5).

These rural communities share a key characteristic: their proximity to significant water bodies and their active participation in social enterprises with a tourism focus. This dynamic strengthens local development and is directly linked to SDGs 6, 8 and 13. The activities carried out in these enterprises (Table 8.6) include ecotourism, guided cultural and nature tours, local gastronomy, artisanal production and lodging services in camping areas and cabins.

Based on the analysis of relationships among the actors involved in these enterprises, there is a notable alignment in the processes aimed at achieving economic improvement while preserving local culture and the environment where tourism activities take place. The synthesis reveals a high degree of community cohesion in protecting the local water resources, which are considered an essential part of the area's appeal to visitors.

This shared recognition of the value of water as a central axis linking economic development, local identity and environmental sustainability positions social entrepreneurship as the core of a complex system. Within this system, various elements—such as the rural community, tourism and water—interact to advance the achievement

Table 8.3 Context-Problem Matrix

		Participants		
		Unitary	Pluralist	Coercive
System	Simple	Simple unitary	Simple pluralist	Simple coercive
	Complex	Complex unitary	Complex pluralist	Complex coercive

Source Jackson (2019)

Table 8.4 Elements of the system under study

Element	Definition
Tourism	Phenomenon generated by the interaction of stakeholders with the environment (tourist areas) and the acquisition of goods or services
Social entrepreneurship	Phenomenon whose main objective is to address social or environmental problems, in addition to generating economic benefits
Rural area	Geographic area characterized by low population density and a close relationship with agriculture, livestock, fishing, or forestry
SDGs 6, 8 and 13	Set of related sustainability goals
Green Infrastructure Regions	Territorial management instrument implemented in Mexico City
Mexico City	Federal Entity of Mexico
Water management	Set of actions and strategies used to conserve and distribute water resources efficiently and sustainably
Boroughs	Territorial demarcation
Landscape	Visible expanse of land encompassing everything that can be observed in a particular place, whether natural or human made
Biodiversity	Variety of life forms on Earth
Rural community	Host community residing within the rural areas of Mexico City
Sustainability	Responsible and balanced management of natural resources, social and economic to ensure that current development does not compromise the well-being of future generations
Economic development	The process by which an economy increases its production of goods and services to improve the standard of living of its population

Source Own elaboration

of the SDGs. Figure 8.5 summarizes these relationships, illustrating how social entrepreneurship becomes a strategic node connecting local action with global sustainability commitments.

When entrepreneurs share distinct values and beliefs but belong to and reside in the same rural community, agreements often emerge that reflect a commitment to environmental stewardship. These agreements include recognizing water as a fundamental resource for maintaining the persistence of these ecosystems and, indeed, as an inherent part of their culture and identity. Furthermore, this methodological approach identified the positive influence of social entrepreneurship and water management on tourism dynamics. This demonstrates that these enterprises foster improvements in the relationships within the system and its components, helping to identify the challenges faced by the system.

Synthesizing the interpretations of actors and system relationships, a complex system is characterized in which decision-making relationships fluctuate between unitary and pluralist approaches. In some cases, enterprise actors align in processes to improve the local economy, preserve culture, and protect the environment within the context of tourism dynamics. Conversely, when differences in values and beliefs

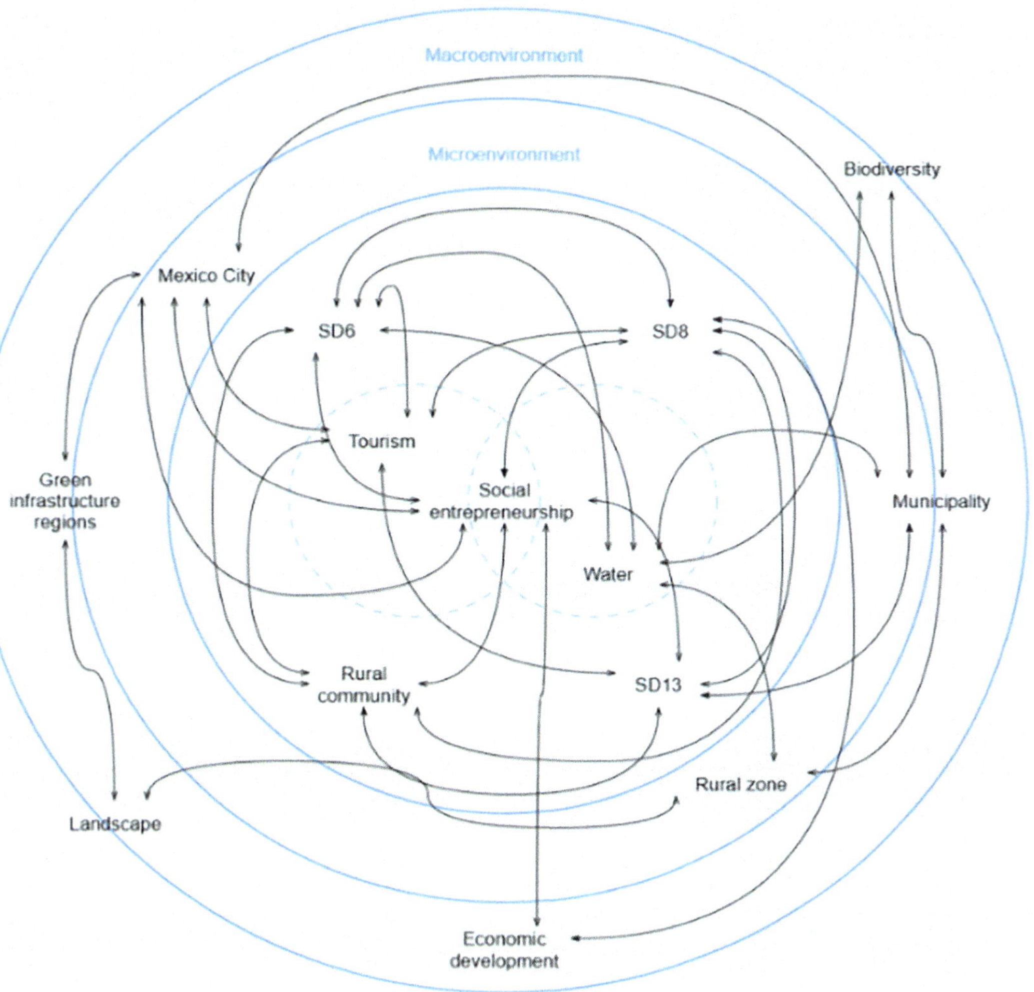

Fig. 8.4 Definition of the system under study. *Source* Own elaboration

exist among actors, agreements emerge that demonstrate a shared commitment to environmental stewardship, integrating ecosystems and recognizing water as a fundamental resource and an essential part of their cultural identity.

8.4 Conclusions

Water management in the rural communities of the boroughs Álvaro Obregón, Cuajimalpa de Morelos, Magdalena Contreras, Milpa Alta, Tláhuac, Tlalpan and Xochimilco in Mexico City plays a significant role in sustainability and overall well-being, especially within the framework of sustainable development. The convergence of tourism and social entrepreneurship offers a unique opportunity to address the challenges associated with sustainable water management, creating synergy between social entrepreneurship and tourism that paves the way for new perspectives in developing inclusive and sustainable business models.

The integration of social and environmental responsibility principles into social tourism entrepreneurship can positively influence the awareness and environmental education for tourists, fostering a culture of responsible tourism and promoting individual or collective actions to protect water resources. This can improve equi-

Table 8.5 Rural communities belonging to the boroughs within the system under study

Borough	Rural community	Proximity to water bodies	Type of social tourism enterprise	Relation to the SDGs
Álvaro Obregón	San Bartolo Ameyalco	Magdalena River and Desierto de los Leones National Park	Ecotourism, guided tours (cultural and natural), local gastronomy, handicrafts, camping areas, and cabins	6, 8, and 13
Cuajimalpa	San Pablo Chimalpa	Magdalena River		
Magdalena Contreras	San Nicolás Totolapan	Magdalena River and Desierto de los Leones National Park		
Milpa Alta:	San Pedro Atocpan	Magdalena River and Desierto de los Leones National Park		
Tláhuac:	1. Santa Catarina Yecahuitzotl 2. San Andrés Mixquic	Lake of the Aztec Kings		
Tlalpan:	1. San Miguel Topilejo 2. San Andrés Totoltepec	Magdalena River and Desierto de los Leones National Park		
Xochimilco	1. San Gregorio Atlapulco 2. San Luis Tlaxialtemalco	Lake Xochimilco		

Source Own elaboration with data from the Government of Mexico City (2023), PAOT (2024a), PAOT (2024b)

Table 8.6 Types of tourism products and services identified in social enterprises

Type of social tourism enterprise	Description
Ecotourism	Activity carried out by community entrepreneurs aimed at encouraging visitors or tourists to appreciate nature responsibly, minimizing negative impacts on ecosystems and promoting environmental and cultural awareness within the locality
Guided tours (cultural and natural)	Conducted by community members who guide visitors through natural or cultural sites to provide an enriching and educational experience, allowing them to learn about the history, culture, biodiversity and the importance of preserving their community
Local gastronomy	Composed of community cooks who seek to empower their identity through local cuisine and the consumption of regional foods
Handicrafts	Artisans or artisan groups who create objects or products using manual skills and traditional techniques, often passed down from generation to generation. These enterprises aim to foster cultural identity and the community's heritage
Camping areas and cabins	Basic facilities or spaces provided by the community for visitors or tourists to set up tents, with areas for campfires, restrooms and sometimes additional services such as drinking water, showers and electricity. These spaces are maintained by the entrepreneurs. Cabins, on the other hand, are small, rustic structures that offer a more comfortable shelter for visitors

Source Own elaboration

table access to water, especially for rural communities that face greater challenges regarding water availability and quality.

The systemic approach applied with TSI reveals the importance of social entrepreneurship in rural areas of Mexico City to improve water management and highlight the products and services offered by social enterprises. It also provides a valuable platform for understanding and addressing the complex challenges associated with water management in these rural communities. This systemic approach further emphasizes effective collaboration among various stakeholders to design solutions that promote sustainability and well-being in these areas.

In this regard, to promote sustainable water management and generate tangible positive impacts in rural communities, as well as to

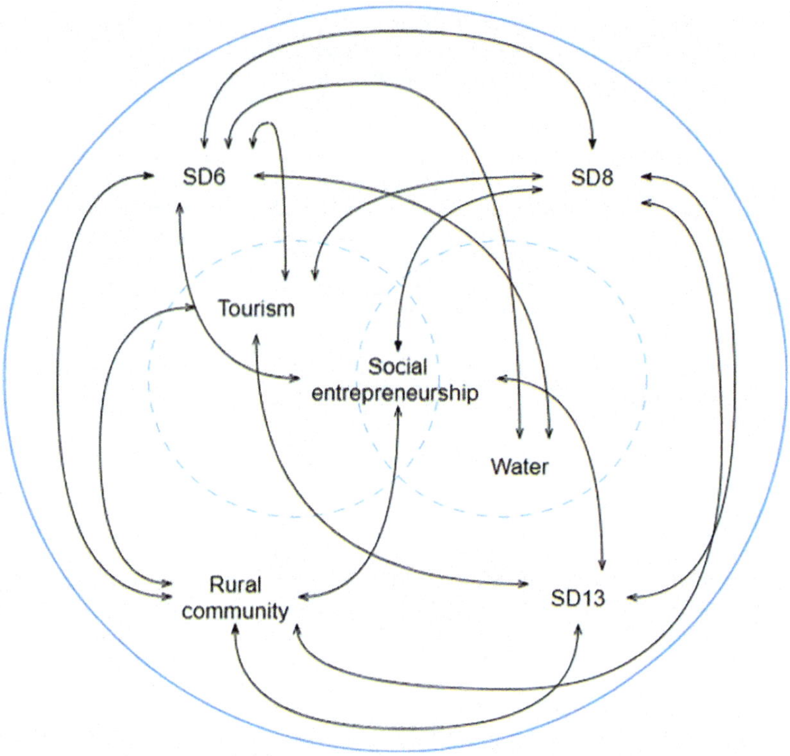

Fig. 8.5 System in focus. *Source* Own elaboration

contribute to the achievement of SDGs 6, 8 and 13, it is essential to foster multi-stakeholder collaboration. This involves creating strategic alliances at both micro and macro levels (government institutions, non-governmental organizations and the rural community) to develop and implement innovative and sustainable solutions that address both current and future water-related challenges.

The interdependence and complexity of the system in focus are highlighted, where water management, tourism and social entrepreneurship are interconnected. This holistic perspective recognizes the need to address not only technical and economic aspects but also the social, cultural and environmental factors that influence their effectiveness and sustainability in the short, medium, and long term. Nevertheless, water management is conditioned by a series of social factors, such as community cohesion, environmental commitment and the integration of cultural values. This underscores the importance of a systemic and collaborative approach that takes into account the diversity of context and the reality of rural communities.

According to the principles of TSI and the findings obtained, for future research with a systemic approach, it is suggested to apply Organizational Cybernetics and Soft Systems in a complementary manner, based on the Systems Methodology Framework, in order to unfold, interpret and manage the complexity of the system under study and to generate constructs for the transformation and improvement of reality.

Acknowledgements This research is supported by the Instituto Politécnico Nacional through transdisciplinary research project SIP3420, modules 20254027 and 20254253, by the Secretaría de Investigación y Posgrado and the Secretaría de Ciencia, Humanidades, Tecnología e Innovación.

References

Aman S, Hassan N, Khattak M, Moustafa M, Fakhri M, Ahmad Z (2021) Impact of tourist's environmental awareness on pro-environmental behavior with the mediating effect of tourist's environmental concern and moderating effect of tourist's environmental attachment. Sustainability 13(23):12998. https://doi.org/10.3390/su132312998

Babaei S, Ghazavi R, Erfanian M (2018) Urban flood simulation and prioritization of critical urban sub-catchments using SWMM model and PROMETHEE II approach. Phys Chem Earth Parts A/B/C 105:3–11. https://doi.org/10.1016/j.pce.2018.02.002

Boroomandnia A, Bozorg-Haddad O, Bahrami M, Goharian E, Singh VP, Loáiciga HA (2021) Optimizing urban stormwater control strategies and assessing aquifer recharge through drywells in an urban watershed. Hydrogeol J 29(4):1379–1398. https://doi.org/10.1007/s10040-021-02316-0

Bouras C, Kokkinos V, Michos E (2020) An online tool on sustainable water management. Tourism 68(4):450–465. https://doi.org/10.37741/t.68.4.6

Cervantes-Jiménez M, Díaz-Delgado C, González-Sosa E, Ángel Gómez-Albores M, Mastachi-Loza CA (2020) Proposal of a water management sustainability index for the 969 sub-basins of Mexico. J Maps 16(2):432–444. https://doi.org/10.1080/17445647.2020.1763486

Chopra A, Ramachandran P (2021) Understanding water institutions and their impact on the performance of the water sector in India. Water Policy 23(2):466–486. https://doi.org/10.2166/wp.2021.207

Cockburn-Wootten C, McIntosh A (2020) Improving the accessibility of the tourism industry in New Zealand. Sustainability 12(24):10478. https://doi.org/10.3390/su122410478

DENUE (2024) National statistical directory of economic units. https://www.inegi.org.mx/app/mapa/denue/default.aspx

Dwyer L (2022) Tourism contribution to the SDGs: applying a well-being lens. Government of Mexico City. (2023). Sustainable city. https://informedegobierno.cdmx.gob.mx/ejes/02-ciudad-sustentable-2/

Epuran G, Tescașiu B, Tecău AS, Ivasciuc IS, Candrea AN (2021) Permaculture and downshifting-sources of sustainable tourism development in rural areas. Sustainability 13(1):230. https://doi.org/10.3390/su13010230

Grison C, Koop S, Eisenreich S, Hofman J, Chang IS, Wu J, Savic D, Van Leeuwen K (2023) Integrated water resources management in cities in the world: global challenges. Water Resour Manage 37(6–7):2787–2803. https://doi.org/10.1007/s11269-023-03475-3

Gutierrez ELM, Vafadari K (2023) Women in community-involved tourism enterprises: experiences in the Philippines. Asia Pac Viewpoint 64(1):85–97. https://doi.org/10.1111/apv.12361

INEGI (2024) Mexico City: territory, area, climate, and water. National Institute of Statistics, Geography, and Informatics. https://cuentame.inegi.org.mx/monografias/informacion/df/territorio/default.aspx?tema=me&e=09

Jackson MC (2019) Critical systems thinking and the management of complexity. John Wiley & Sons

Khan MR, Khan HU, Lim CK, Tan KL, Ahmed MF (2021) Sustainable tourism policy, destination management and sustainable tourism development: a moderated-mediation model. Sustainability 13(21):12156. https://doi.org/10.3390/su132112156

Kummitha HR, Kolloju N, Jancsik A, Szalók ZC (2021) Can tourism social entrepreneurship organizations contribute to the development of ecotourism and local communities: understanding the perception of local communities. Sustainability 13(19):11031. https://doi.org/10.3390/su131911031

Li X, Abbas J, Dongling W, Baig NUA, Zhang R (2022) From cultural tourism to social entrepreneurship: role of social value creation for environmental sustainability. Front Psychol 13. https://doi.org/10.3389/fpsyg.2022.925768

Liu H, Jiang Y, Zhu H, Chen Y, Lyu W, Luo W, Yao W (2021) Analysis of water resource management in tourism in China using a coupling degree model. Water Policy 23(3):765–782. https://doi.org/10.2166/wp.2021.155

Nhemachena C, Nhamo L, Matchaya G, Nhemachena CR, Muchara B, Karuaihe ST, Mpandeli S (2020) Climate change impacts on water and agriculture sectors in Southern Africa: threats and opportunities for sustainable development. Water 12(10):2673. https://doi.org/10.3390/w12102673

PAOT (2024a) Delegational urban development program of Álvaro Obregón. https://paot.org.mx/centro/programas/delegacion/alvaro01.html

PAOT (2024b) Delegational urban development program of Cuajimalpa de Morelos. https://paot.org.mx/centro/programas/delegacion/cuajima.html

Papaluca O, Tani M, Troise C (2020) Entrepreneurship and sustainability in tourism: an interpretative model. J Mgmt Sustain 10:38

Paudyal NP (2019) Tourism education and training for entrepreneurship development. Third Pole J Geogr Educ, 97–110

Picciotto L, Ummarino I (2022) Entrepreneurship of people with disabilities and accessible tourism in the non-hotel accommodation sector in Italy (pp 353–374). https://doi.org/10.31410/tmt.2022-2023.353

Popkova EG, Bogoviz AV, Lobova SV, Delo P, Sergi BS, Yankovskaya VV (2023) Global transitions towards social entrepreneurship and sustainable development: a unique post-COVID-19 perspective. Global Trans 5:182–200. https://doi.org/10.1016/j.glt.2023.08.005

Rela IZ, Awang AH, Ramli Z, Md Sum S, Meisanti M (2020) Effects of environmental corporate social responsibility on environmental well-being perception and the mediation role of community resilience. Corp Soc Responsib Environ Manag 27(5):2176–2187. https://doi.org/10.1002/csr.1956

Romão J (2018) Tourism, territory and sustainable development. In: Springer (Vol 28). Springer Singapore. https://doi.org/10.1007/978-981-13-0426-2

Santos E, Moreira J (2021) Social sustainability of water and waste management companies in Portugal. Sustainability 14(1):221. https://doi.org/10.3390/su14010221

SEDEMA (2019) Special green infrastructure program of Mexico City, 7–24

SEDEMA (2024) Green infrastructure program. Secretariat of the environment. https://sedema.cdmx.gob.mx/programas/programa/infraestructura-verde

Sharma V, Bhat DAR (2023) The role of community involvement in sustainable tourism strategies: a social and environmental innovation perspective. Bus Strategy Dev 6(2):119–127. https://doi.org/10.1002/bsd2.227

Skivko M (2021) Digital technologies, social entrepreneurship and governance for sustainable development. Res Soc Change 13(1):165–173. https://doi.org/10.2478/rsc-2021-0016

Tleuberdinova A, Shayekina Z, Salauatova D, Pratt S (2021) Macro-economic factors influencing tourism entrepreneurship: the case of Kazakhstan. J Entrepreneurship 30(1):179–209. https://doi.org/10.1177/0971355720981431

UN (2024a) What are the sustainable development goals? https://www.un.org/sustainabledevelopment/es/development-agenda/#:~:text=Los Objetivos de Desarrollo Sostenible (ODS) constituyen un llamamiento universal, personas en todo el mundo

UN (2024b) Goal 8: decent work and economic growth. https://www.un.org/sustainabledevelopment/es/economic-growth/

UN (2024c) Sustainable development goals. https://www.un.org/sustainabledevelopment/es/objetivos-de-desarrollo-sostenible/

Vuković M (2022) The application of GIS in sustainable tourism management. Econ Sustain Dev 6(2):53–62. https://doi.org/10.5937/ESD2202053V

Wu X, Liang X (2023) Tourism development level and tourism eco-efficiency: exploring the role of environmental regulations in sustainable development. Sustain Dev 31(4):2863–2873. https://doi.org/10.1002/sd.2555

Xiao D, Su J (2022) Role of technological innovation in achieving social and environmental sustainability: mediating roles of organizational innovation and digital entrepreneurship. Front Public Health 10(March):850172. https://doi.org/10.3389/fpubh.2022.850172

Zamanzad-Ghavidel S, Sobhani R, Etaei S, Hosseini Z, Montaseri M (2021) Development of hydro-social-economic-environmental sustainability index (HSEESI) in integrated water resources management. Environ Monit Assess 193(8):463. https://doi.org/10.1007/s10661-021-09129-4

9 Pantry Shopping as a Means of Empowering Women Responsibly

María Angélica Cruz Reyes, Evelia Rojas Alarcón, and Romina Castillo Malagón

Abstract

The objectives of this research are to identify the factors prioritized by women who are responsible for purchasing household groceries; to analyze women's discernment regarding responsible empowerment when buying processed foods for the family pantry; and finally, to determine the degree of relationship between purchase decisions and the sustainable value of agri-food companies. The study employs structural equation modeling using *Partial Least Squares*, and is grounded in the natural resource-based view of the firm, purchasing rationality, and economic empowerment theories. The aim is to contribute to the literature on empowerment from an administration perspective, incorporating statistical analysis tools. Among the main findings, more than 50% of surveyed women are willing to pay a premium for foods with longer shelf life, with taste being the most important factor in their purchasing decisions. The respondents perceive that by choosing to buy processed products, they do contribute to creating sustainable value for organizations.

Keywords

Women's empowerment · Sustainable value · Agri-food industry · Purchasing decision

M. A. C. Reyes (✉) · E. R. Alarcón · R. C. Malagón
Instituto Politécnico Nacional (IPN), CDMX, Mexico
e-mail: macruzr@ipn.mx

E. R. Alarcón
e-mail: erojasa@ipn.mx

R. C. Malagón
e-mail: rocastillom@ipn.mx

9.1 Introduction

Gender studies related to the environment and sustainability have emerged as an analytical tool for equality and economic empowerment. However, little research has been conducted in the fields of Administration, Accounting, and related areas. In the context of sustainable development, gender inequalities are increasingly being highlighted in order to develop mechanisms and strategies for economic empowerment.

For example, studies in Chile suggest that the aspects differentiating gender roles include the sexual division of labor, such as domestic tasks and childcare, which are primarily the responsibility of women, while men are mainly seen as providers for the household (Céspedes 2002). In other words, there is an asymmetrical relationship between men and women.

Among the goals of the 17 Sustainable Development Goals (SDGs) is the elimination of gender inequality. Women's economic empowerment is a prerequisite for achieving this, through the eradication of discrimination and violence against women and girls (UN Women 2017).

P. Pérez-Hernandez and R. Huerta Cuervo (eds.), *Social Well-Being and the Challenges in the Sustainable Development Goals*, Sustainable Development Goals Series,
https://doi.org/10.1007/978-3-032-14294-8_8

Accordingly, the analysis of gender equality begins with the recognition that women have historically been discriminated against based on age, race, ethnicity, sexual orientation, or socioeconomic status, despite the fact that the right to equality and non-discrimination is enshrined in the international human rights protection system (United Nations [UN] Women's Section 2015).

The UN's strategy for women's economic empowerment can represent an opportunity not only to break the glass ceiling but also to increase organizational effectiveness, reduce vulnerability and poverty. This research focuses on women responsible for grocery shopping and meal preparation, as they possess the sensitivity to drive the transition toward sustainability and gradually move away from the traditional model of the "male provider" and the "female homemaker."

To provide context, globally, the employment rate for men is 72.2%, while for women it is 47.1%, with women earning less on average. In Latin America and the Caribbean, the likelihood of women being employed in informal sectors exceeds 50.0%, increasing their vulnerability due to job insecurity (United Nations [UN] 2017). In this regard, Guzmán (2002), citing Godoy and Mauro (2001), explained that changes (albeit slow) in work organization are due to shifts in conditions and positions in the labor market. It also describes that when both partners in a couple are employed, it is difficult for either of them to permanently take on household tasks, which leads to a renegotiation of the division of roles within the couple, with both becoming providers for the household.

Thus, in the twenty-first century, women have gradually entered the workforce, moving from rural to urban areas, driving changes in roles, transforming their environment, and creating trends such as those in consumption. At the same time, they are often cited as the main factor in the so-called crisis of masculinity. On the other hand, Girón (2012: 45) noted that "women's equality becomes fundamental both in socialism, to increase productivity, and in capitalism, to boost household income and consumption."

The objectives of this research are: 1. To identify the key factors prioritized by women who make decisions regarding the purchase of household groceries; 2. To analyze women's discernment regarding responsible empowerment when purchasing processed foods for the household pantry; and 3. To determine the degree of correlation between purchasing decisions and the sustainable value of agri-food companies.

The research hypothesis is that the decision-making process for household grocery purchases involves factors of control and commitment that empower women.

The hypotheses of the research questions are:

H_1 The main factor in the purchasing decisions of women responsible for household groceries is price.

H_2 When women choose among different processed food alternatives, they drive trends that impact the production and marketing strategies of agri-food companies.

H_3 There is a strong and direct relationship between the purchasing decisions of women responsible for household groceries and the sustainable value of agri-food companies.

To achieve these objectives, a literature review was first conducted to identify studies that provide insights into empowerment and gender (*empowerment AND gender*) in the *Web of Science* database. As of January 30, 2024, 7,940 research articles were found on these topics. However, when limited by field, only 267 articles were found in business journals, and 419 in the field of education. In the Springer database, 123 documents were identified using the keywords women's empowerment and female consumers; when the term gender was included, the search narrowed to 79 results. This indicates that studies on empowerment and gender are rarely addressed in the fields of management, business leadership, administration, or business.

The theoretical framework for this research is based on the resource-based view of the firm, purchasing rationality, and sustainable value. Subsequently, the empirical research instrument was designed and analyzed using the structural equation modeling approach with *Partial Least Squares (Smart PLS software)* to determine the weight of each observable variable and the R^2 coefficient to test the hypotheses.

9.2 Women's Empowerment

Research on women's empowerment and sustainable development has been complemented by proposals such as: Feminism and Social Change in Latin America and the Caribbean (2012); Gender, Equity, and Citizenship (2005); Gender and Environment (2003); Power and Women's Empowerment (1997); as well as publications by UN Women at the international level and more than 60 gender working papers by Inmujeres Mexico. These documents highlight the importance of the topic for achieving sustainable development.

Meanwhile, in the field of business and economics, the critical management perspective focuses on the discourses used to legitimize teamwork, which foster a functionalist view of the firm. Alongside gender-based analysis, this has led to greater understanding of workplace and organizational relationships, exposing gender hierarchies and discrimination. Today, concepts that previously existed but were not openly discussed—such as sexual harassment, the glass ceiling, work-family balance, masculinities, and bodies—are now visible. In other words, organizational processes are analyzed with sensitivity to forms of work and organization (Adler et al. 2008).

Adler et al. (2008) also made a critique of labor markets in the context of capitalism, explaining that it is the means by which the capitalist system exerts power over workers and structures the division of labor, thereby distinguishing between the more and less exploited. Moreover, diversity factors such as race and gender are used to divide people. They argue that the consumer market is the means to create demand to satisfy desires generated by advertising.

Through strategies for gender equality and women's empowerment, such as assigning market value to household tasks, it is possible to move from the discrimination of vulnerable jobs to recognizing them as delicate and specialized due to the understanding of the skills required to perform them, or to assess purchasing power by the decision-making capacity involved in grocery shopping for the household.

Lamas (2005: 102–103), for her part, highlighted the multifaceted role of women, which makes it possible to recognize women's contributions to regional well-being simply by representing more than 50% of the world's population and by providing household services, understood as the upbringing and care of children, the elderly, the sick, food preparation, clothing care, grocery shopping, and volunteer work.

Guzmán (2002), from the Economic Commission for Latin America and the Caribbean [ECLAC], referenced Godoy and Mauro (2001), who argued that changes in the organization of work are driven by shifts in conditions and positions within the labor market. They also illustrated that when both partners in a couple are employed, they renegotiate household roles, and both become providers.

Regarding the challenges of the twenty-first century, employment among men and women is fundamental to achieving sustainable development. From the perspective of stratification studies, a person's integration into the productive structure is considered a relationship between the individual and the social structure (Sautu 1992, cited by Fraga and Riveiro 2011). In other words, occupation is one of the variables in these studies, aiming to identify formal tasks and skills derived from education and work experience. However, one of the main criticisms of this approach is the limited attention given to gender perspective.

According to Chávez (2006: 9), this is important because it is a theoretical framework that allows for the analysis, interpretation, and explanation of the social relationships established between men and women at a given historical moment. Within this approach, gender must be analyzed comprehensively, as women have been subjected to vulnerability, abuse, and discrimination not only because of their sex but also due to historical social constructs.

Here, the "housewife" is defined as a self-employed worker in the domestic sector responsible for managing and producing the household's domestic work. Torres (1989) identified how women perform two basic functions: management and execution, and how they do not have assistance to carry out these tasks.

This is evident in the activities that women have traditionally performed in assembling the family pantry, which are the same activities that set the stage for the act of freely deciding on purchases. As described by Freire (1970), this act, when carried out as a commitment to transformation, initiates the process of liberation of the oppressed; because learning comes from the oppressed and from those who are truly in solidarity with them, which is the true generosity needed to transform the world. The same author states that, to overcome oppression, people must first critically recognize its causes so that, through transformative action, they can create a new situation—one that enables the pursuit of a more complete humanity (Freire 1970: 47; Colomer and Peiró 2002).

On the other hand, from the perspective of planned social change, Young (1970), using a comprehensive framework of involvement for development, specified that empowerment involves changing the processes and structures that reproduce women's subordinate position as a gender. That is, for change to promote equality and not perpetuate male hegemony, a fundamental element is decision-making power from the planning stage.

Agha et al. (2018) argue that textbooks continue to reinforce family roles in which women are responsible for domestic tasks and are depicted as secondary citizens in society.

Meanwhile, León (1997, pp. 14–20) defines empowerment as a process of overcoming gender inequality. From the perspective of existing power relations, such empowerment seeks to gain greater control over sources of power. When the author cites Foucault and Lauretis, it is to explain that the meaning of power is multidimensional, as it is "the confrontation between opposites as something structural," being both individual and collective-political; it can be static (lack of power) but also dynamic; the process is aimed at decision-making in the face of problems or conflicts.

With regard to women's economic empowerment within the framework of policies to achieve the United Nations Millennium Development Goals, it was defined in terms of five components: (1) Self-worth; (2) The right to have and choose options; (3) The right to access opportunities and resources; (4) The right to have control over their own lives, both at home and outside it; and (5) The ability to influence social change to create a more just social and economic order at the national and international levels (Cordourier 2015).

Based on this argument, the empowerment of a vulnerable person can be understood as a process of change that moves from insecurity to security, from dependence and marginalization to independence, decision-making, and self-esteem. For this reason, education is fundamental. However, as an example, Elkhweaky et al. (2023), in the context of studies on business leadership, note that the challenge is even greater, as women continue to face gender biases and stereotypes, work-life balance issues, and a lack of equity programs and policies that would help reduce the gender gap.

Similarly, Gupta and Roy (2023) concluded that economic empowerment is important for women's decision-making capacity; however, intervention programs must be designed for such empowerment, with an emphasis on education.

9.3 Sustainable Value

To outline the construct of "sustainable value," Hart (1995) was referenced, specifically "A natural-resource-based view of the firm," which proposes studying the relationship between the environment (ecological) and the company. The author argued that the strategic capacity of an organization should focus on: the prevention of contamination, product management, and sustainable development, whose main drivers are the minimization of waste emissions, the reduction of product life cycle costs, and the mitigation of the environmental burden associated with the company's growth and development, respectively. To achieve this, company resources must be focused on continuous improvement, stakeholder integration, and the establishment of a shared vision.

In 2003, Hart and Milstein proposed that a company should focus on "creating sustainable value." Among the drivers to be considered for this purpose are both population and consumption. In the case of population, this involves strategies for the company's future and growth, but based on actions aimed at minimizing inequality and poverty (Hart and Milstein 2003).

In other words, it is about identifying new business opportunities at the base of the population pyramid, while for consumption, the focus should be on products that minimize both pollution and the company's costs and expenses. They also noted that the impact of the triple bottom line (social, environmental, and economic) of the business system should be assessed, as both positive and negative effects are observed throughout the supply chain or in the final use of the product or service. In this way, the sustainable value model is designed by combining the company's internal and external factors, both in the present and the future (see Fig. 9.1) (Hart and Milstein 2003).

Figure 9.1 presents the classification of drivers of sustainable value. In the upper right section, the strategic proposal for growth is shown, which is based on the population pyramid so that the organization not only grows for the future but also works to minimize poverty and inequality.

Subsequently, Hart (2007: 198) added to his initial proposal the element of business model innovation for the base of the population pyramid, emphasizing the need to create surplus and generate income through the consumption of low-income individuals, as these populations are often underserved by companies, meaning there are gaps in the services provided to them. Hart (2007) also pointed out that poor people are subject to discrimination and are vulnerable due to a lack of favorable economic conditions to meet their basic needs; in this regard, women are distinguished by these same characteristics.

9.4 Product Purchase Factors

There has been a significant proliferation of research explaining the relationships between consumers and the products or services that can meet their needs. Specifically, behavioral economics studies consumer attitudes and behavior, explain consumption patterns based on life cycles and lifestyles, brand loyalty as a function of its characteristics and consumer behavior, and discuss the links between behavioral economics and consumer decision-making processes (Ratchford 2001; DiClemente and Hantula 2003; Montgomery 2011).

These authors state that the choice of goods acquired by consumers can be explained under different assumptions:

1. Marginal utility, given the price determined by the quantities demanded and supplied, based on the logic of the utility (satisfaction) derived from these goods.
2. Regulation, since even in adverse environmental conditions, consumers primarily purchase goods necessary for survival, such

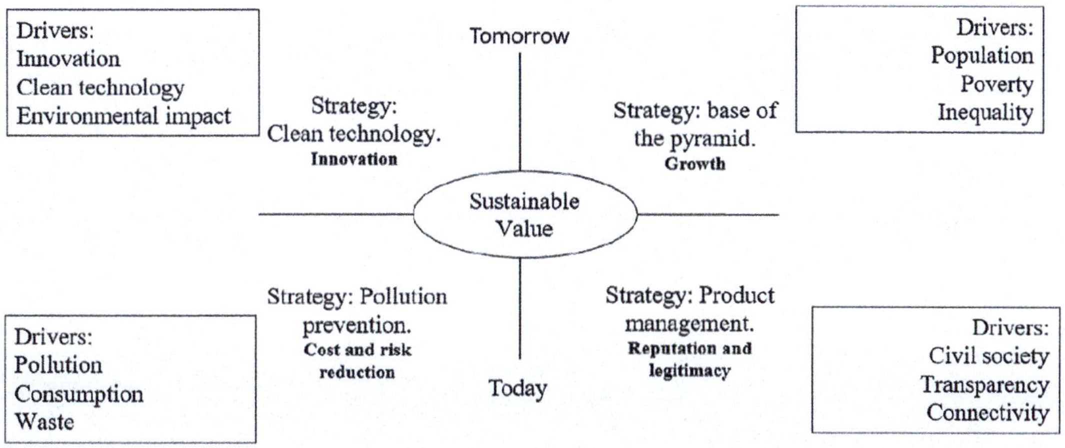

Fig. 9.1 Drivers of sustainable value. Prepared with reference to Hart and Milstein (2003) and Hart (2007)

as food, followed by habits, lifestyles, or consumption patterns, and finally by environmental influences such as the availability of substitutes, accessibility, and supply.

However, it was Lancaster (1966), with the New Consumer Theory, who broke with the traditional logic that goods themselves provide utility. Lancaster posited that it is the characteristics of products that generate utility, since "the only property goods share is that they are simply goods" (p. 132). For example, a car is preferred by the consumer not for itself, but for its color, spaciousness, or features, as consumption is an activity in which goods or sets of goods are inputs that, in turn, comprise a collection of characteristics that indirectly contribute to consumer satisfaction and preference.

Boluda and Capilla (2017), influenced by Shepherd and Stockley (1987), Shepherd (1990), Steptoe et al. (1995), and Roininen et al. (1999), conducted studies to relate consumer attitudes to purchase choices, under the theory of planned behavior. They concluded that not only attitudes and beliefs influence food choices, but demographic, environmental, health, mood, convenience, price, ethical, and taste factors also impact decision-making. These authors based their work on a model with four categories to identify attitudes toward functional foods and consumer behavior, considering barriers and motivators. The categories are: reward (health, mood, well-being); necessity (to improve health); trust (for promised benefits); and safety (perceived safety).

Beardsworth et al. (2002), Baker and Wardle (2003), Urala and Lähteenmäki (2003), and Ridler and Ridler (2011) also inspired Boluda and Capilla (2017) to determine that the relationship between healthy lifestyles and attitudes toward functional foods is moderated differently according to gender.

9.5 Purchase Decisions and Economic Empowerment

From a Keynesian perspective, Girón (2012) explained that "women are creators of monetary wealth through their participation in paid work and participate as consumers by stimulating the demand and consumption within the economic process" (p. 45). He also described that "gender equality is fundamental in capitalism to increase household income and consumption" (2012: 45).

Below, Fig. 9.2 is presented, which illustrates the effects of women's empowerment through purchasing decisions, viewed as a hegemonically basic task of women within the household.

Figure 9.2 shows how an endogenous basic household task leads to an exogenous one when making a purchase decision. According to Girón

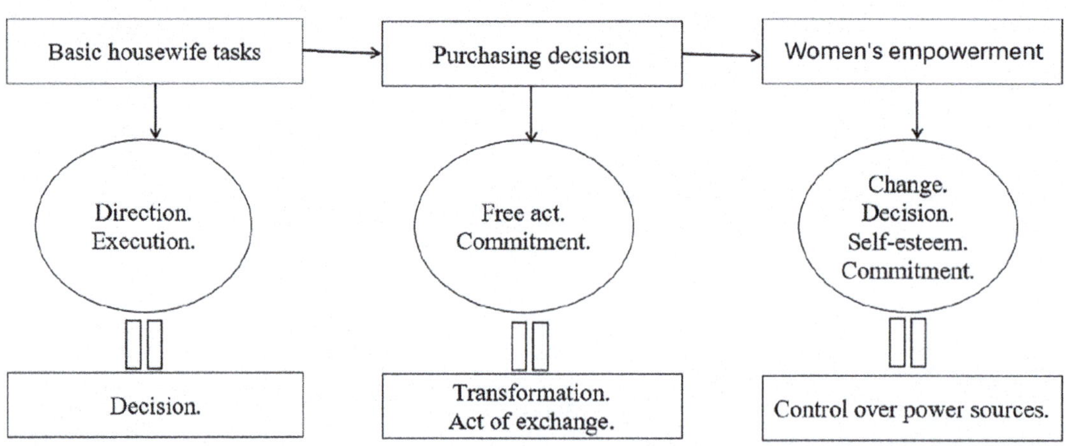

Fig. 9.2 Relationship between basic household tasks, purchasing decisions, and women's empowerment. *Source* Prepared with reference to Freire (1970), Young (1970), León (1977), Torres (1989), and Lamas (2005)

(2012), this in turn results in women's empowerment. Women who perform household activities do not necessarily know the functions of management, but at home, activities such as direction and administrative control are carried out, characterized by commitment and freedom for change.

9.6 Agri-Food Industry in Mexico

The agroindustry depends on raw materials from the primary sector; however, agriculture faces various risks due to the impacts of climate change. According to Bárcena (2020), rising temperatures, extreme natural events, or increased fertilization due to higher CO_2 levels pose high-impact risks for agriculture, affecting both production and food quality, leading to higher prices, greater vulnerability among populations facing food insecurity, and increased income inequality.

According to Flores et al. (1986), the agroindustry is a social production process that prepares, preserves, and/or transforms inputs originating from the primary production sector, diversifying consumption through the creation of new products. Meanwhile, the agri-food industry is a social process that, like the agroindustry, aims to transform natural foods or inputs from agriculture, livestock, fishing, and forestry into processed foods for the end consumer, who uses them to meet human or animal nutritional needs.

The Food and Agriculture Organization of the United Nations (FAO 2013) identified three factors classified as megatrends for the agroindustry: 1. Convenience; 2. Consumption patterns; and 3. Women's participation in paid labor activities.

In 2017, the FAO also identified 15 trends for the future of food and agriculture, including "changes in food systems" driven by consumption habits such as those between rural and urban areas, between purchasing power and monetary scarcity, between urban population growth and land use change, and even by advertising or affordability and availability. As a result, there is an increasing need for processed foods that can be easily transported, stored, distributed, and marketed to reach consumers (Bárcena 2020).

It is worth noting that, according to the FAO, by 2025, food demand will increase by 70 to 90%, and by 2050, global population dynamics are expected to change radically, reaching nearly 9.7 billion people, with the greatest growth in developing countries. Food production and distribution will be among the main challenges to address (FAO 2017).

The importance of the food industry throughout its value chain can be explained by each of its links, starting with seed supply, followed by agriculture, processing, and the direct and indirect distribution of processed foods. The goal is to produce safe foods with longer shelf life and greater availability, that is, foods that comply with food safety standards.

In Mexico, according to the Secretariat of Agriculture, Livestock, Rural Development, Fisheries, and Food (SAGARPA), the agri-food sector is strategic for the country's economy and development. It comprises subsectors such as agriculture, livestock, fisheries, aquaculture, and agroindustry, which together contribute more than 8.2% to the Gross Domestic Product (GDP). Internationally, Mexico ranks 12th out of 194 countries as a food producer (INEGI 2023b), with the United States of America (USA) being the main consumer of Mexican foods. Likewise, the food industry contributes to the industrial activity index and thus to the generation of formal employment (see Graph 9.1).

Notes:

1. Preliminary figures. Due to the estimation method, the series may be modified as new information becomes available.
2. The seasonally adjusted series for total Industrial Production and its components (Mining; Generation, transmission, and distribution of electricity, water supply, and gas through pipelines to the final consumer; Construction; and Manufacturing Industry) are calculated independently.

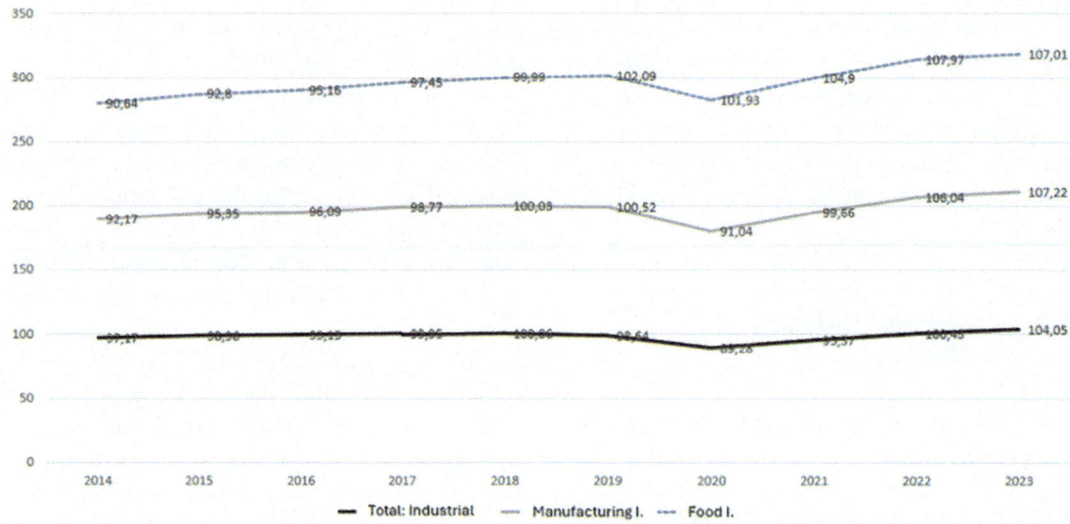

Graph 9.1 Comparison of the average index from 2014–2023 for total industrial, manufacturing, and food industry activity. *Source* Prepared based on INEGI (2023a). Seasonally adjusted series, Index Base 2018 = 100

According to Graph 9.1, the index of food industry activity is above both the manufacturing and the overall, food industry was the least impacted despite the economic and productivity crisis caused by the COVID-19 health emergency.

However, according to a study conducted by the FAO (2022), the COVID-19 pandemic and the ensuing economic crisis have had a negative impact on families' ability to pay for food and basic services, especially in countries with wide inequality gaps and high levels of poverty. In Mexico, the agribusiness sector is significant, as food processing and its characteristics can help reduce food poverty. Although food insecurity decreased from 26.7% in 2018 to 26.1% in 2021 (FAO 2022), it remains high. It is noteworthy that Mexican families allocate the majority of their income to food, beverages, and tobacco (37.7%), while only 3.4% is spent on health (INEGI 2023b).

9.7 Method

To achieve the following objectives: 1. Identify the elements prioritized by women who make decisions regarding household grocery purchases, 2. Analyze women's discernment regarding responsible empowerment when purchasing processed foods for the household pantry, and 3. Identify the degree of correlation between purchasing decisions and the sustainable value of agri-food companies.

The sample selected according to the research interest, in the first stage, consisted of 354 observations, the majority being women from Mexico City, with a smaller percentage from the State of Mexico. The age distribution was as follows: 21 to 30 years, 15.5%; 31 to 40 years, 26.8%; 41 to 50 years, 31.4%; 51 to 60 years, 18.4%; and over 60 years, 7.9%.

35.7% had completed upper secondary education (high school), and 28.3% had completed higher education. Of all the questionnaires distributed, 96.9% of respondents identified as housewives, while the remaining 3.1% were men.

It is important to note that of all valid questionnaires, 43.5% of women were exclusively housewives, while the rest also engaged in additional activities such as owning a business, working in the formal sector, or participating in volunteer work.

In the second stage of the research, the sample consisted of women from Mexico City, with the same characteristics as the first instrument, taking into account the research hypotheses. 77.0% of the questionnaires were administered to female shoppers living in Mexico City (CDMX), and 23.0%

to those from the State of Mexico (suburban areas adjacent to CDMX). This instrument measured the relationship between purchasing power and the sustainable value of agri-food companies, considering product quality and environmental care as variables for sustainable value, and nutrition information (according to product labels) and satisfaction perspective (according to the women interviewed) as variables for purchasing power.

The instrument was validated using expert judgment, and the language obtained a Cronbach's Alpha index of 0.841, indicating high reliability. It was administered to 354 women responsible for purchasing household groceries.

The research assumption is that the decision-making process for household grocery purchases includes factors of control and commitment that empower women.

The research hypotheses are as follows:

H_1 The main factor in the purchasing decision of women responsible for household groceries is price.

H_2 When women choose among different processed food alternatives, they create a trend that impacts the production and marketing strategies of agri-food companies.

H_3 The relationship between the purchasing decision of women responsible for household groceries and the sustainable value of agri-food companies is direct and strong.

Structural modeling of hierarchical and higher-order components was performed using the *partial least squares* (PLS) method. The sample size, according to the structural model, has a significance level and a statistical power of 80%. Hair et al. (2017). The research instrument used a five-point Likert scale, with 1 being "strongly disagree" and 5 "strongly agree" (Fig. 9.3).

The research has a relational scope, as shown in Fig. 9.3. The purchasing decision (PD) is related to women's empowerment (WE) and sustainable value (SV). In turn, the purchasing decision can be influenced by shopping experience and marketing factors.

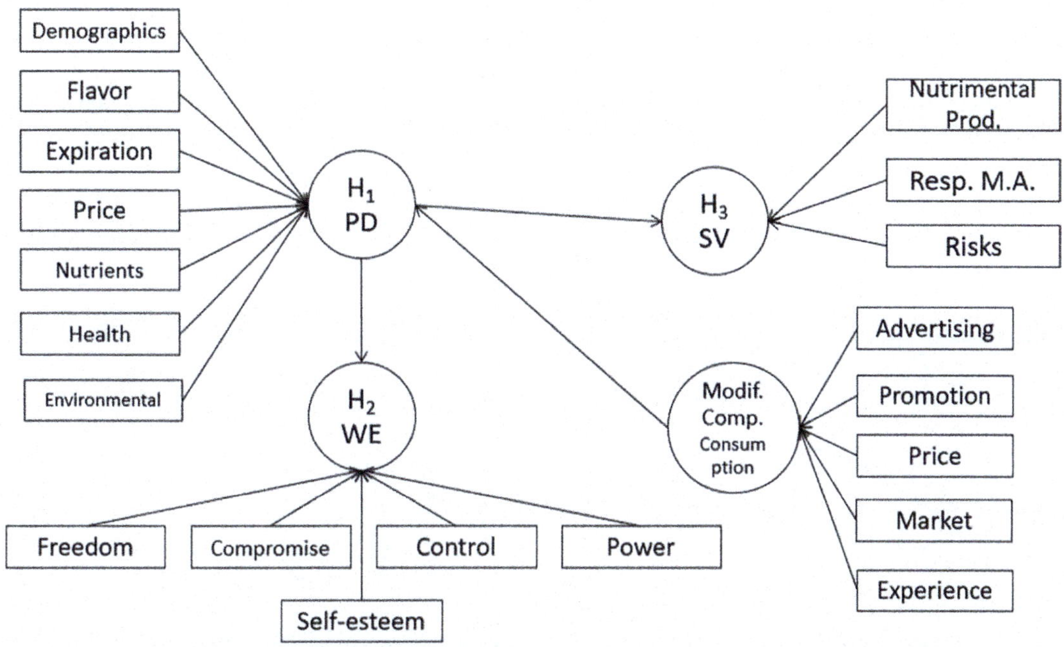

Fig. 9.3 Theoretical formative measurement model: Women–purchasing decision–empowerment. *Source* Compiled by the authors with reference to Hair et al. (2017), FAO (2013), Freire (1970), Young (1970), León (1977), Torres (1989), and Lamas (2005)

Table 9.1 Data analysis of factors, perception, and purchasing decision

	Purchase factors	Purchase decision	Modifies decision	Feeling while buying	An act	Effect of the act
Median	3	4	3	2	2	3
Mode	3	5	3	2	2	5
Median	Health	Nutritional content	Price	Committed	Collective	Social responsibility
Mode	Health	Price	Price	Committed	Collective	Collective

Prepared using Excel based on the questionnaire processed with descriptive statistical analysis

For H_1, Table 9.1 presents the analysis of information regarding the factors influencing the purchasing decision (PD).

According to Table 9.1, the mode was determined based on the most common response, and the median was identified by the response mean between the lowest and highest scores, as H1 was sought. Through descriptive statistical analysis and by considering measures of central tendency, reflecting the frequency of responses, it was found that the main purchasing factor is health. Regarding the purchase decision, nutritional content and price are key; however, based on experience or impulse driven by marketing elements, price remains the prevailing factor.

With respect to external purchasing factors such as demographic, environmental, health, convenience, and mood-related aspects, the purchase decision regarding the characteristics of the food products offered centers on taste, expiration date, price, nutritional content, and product quality.

On the other hand, regarding H_2: whether women feel empowered when purchasing basic food baskets from different alternatives, it was found that the predominant feeling is one of "commitment." The women interviewed described it as a "social responsibility" and an "act of collective action," with "economic empowerment" ranking third. The corresponding histograms are presented below (see Fig. 9.4).

The frequency technique was based on the "n" observations and the cumulative percentage. In a second question in the questionnaire, which aimed to confirm the effect of purchasing on empowerment and whether the purchase decision contributes to national economic growth, the responses indicate that the women interviewed neither fully agreed nor disagreed; only 10% stated that they do feel economically empowered and that they do support the country's economic growth. The information provided by Fig. 9.2 is similar to Table 9.1: the women surveyed reported that they neither fully agreed nor disagreed that

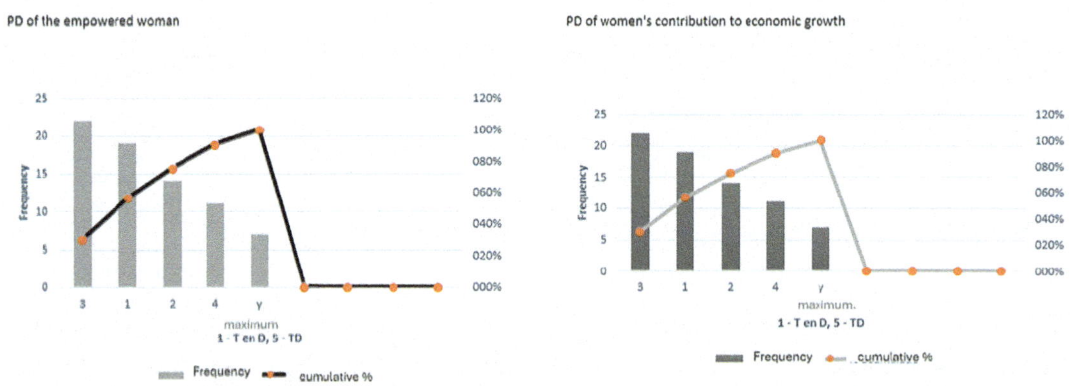

Fig. 9.4 Histogram of purchase decision, empowerment, and economic growth. Prepared in Excel based on the questionnaire processed using frequency analysis

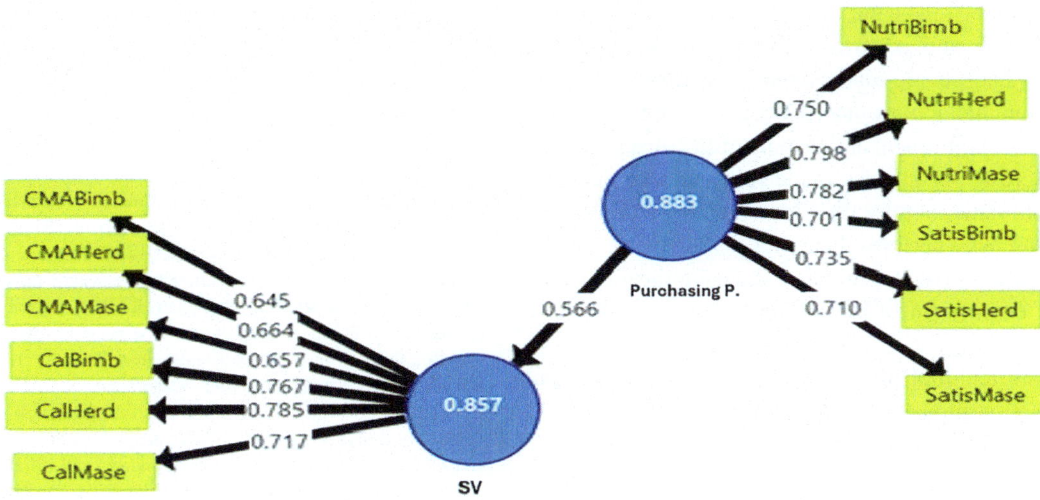

Fig. 9.5 Structural model: relationship between purchasing power and sustainable value. *Source* Own elaboration with reference to Hair et al. (2017)

this action empowers women or supports national growth (Fig. 9.5).

A formative measurement model was used via Smart PLS 2020. The composite reliability was 0.883 for the purchasing power construct and 0.857 for sustainable value, with a correlation of 0.566, indicating a moderate direct relationship. Similarly, the HTMT indicator (heterotrait-monotrait ratio for discriminant validity) was valid, reaching an index of 0.671, which demonstrates strong relationships with its own indicators.

H3 is accepted, as the relationship between the purchase decision of the woman responsible for the family's groceries and the sustainable value of agri-food companies is moderately direct.

9.8 Final Reflections

The research focused on the components of empowerment for gender equality, such as self-worth, freedom of choice, access to opportunities and resources, decision-making, and the ability to influence public, social, and economic spheres, all considered from the perspective of management and organizational strategy.

From the standpoint of organizational strategy, the work draws on Hart and Milstein, who propose that for an entity to endure and generate future opportunities, it must do so through the creation of sustainable value. Notably, one of the drivers of such value is found at the base of the population pyramid, characterized by vulnerability and economic inequality, as is the case for women.

This perspective is relevant as it also encourages the reconsideration of business models using sustainable development as a guiding principle.

In this regard, one of the findings of the research, according to the perceptions of the women surveyed, is that the decision to purchase the basic food basket, which includes processed foods, is based on social factors such as health, nutritional content, and environmental care, as proposed by Hart, who considers these drivers of value generation in a triple bottom line framework.

Regarding women's empowerment, historically, the act of purchasing the family's basic food basket has been the responsibility of the homemaker, which, according to Girón, in an economic system fosters growth, but also, the act of deciding what to buy is one of commitment and freedom.

According to the research findings, it is indeed an act of commitment and social responsibility; however, women do not necessarily feel economically empowered or believe they contribute to the country's growth.

Reflecting on the results, the design of communication and advertising strategies is proposed so that people understand the importance of economic actions, and that by carrying them out with commitment and responsibility, it is possible to create not only economic value but also sustainable value for all stakeholders, while empowering those who perform them, not just women.

This work is original in studying the elements of empowerment in relation to purchasing decisions from a management perspective and in using statistical tools that help elucidate the degrees of relationship between decision-making, empowerment, and sustainable value.

However, it should be noted that these results are an approximation for understanding the decision-making process of women who do not assimilate or understand that the activity of purchasing the basic food basket fosters economic growth, empowers those who make responsible decisions, and at the same time contributes to achieving the goals of sustainable development objectives regarding the empowerment of women and girls to reduce gender inequalities.

Finally, this work would not have been possible without the support of IPN in authorizing SIP research projects.

This chapter was originally written in Spanish and translated using the Springer machine translation service. The authors have reviewed the translation to ensure its accuracy.

References

Adler PS, Forbes LP, Willmott H (2008) Critical management studies. In: Adler PS (ed) The academy of management annals, 119–179. https://www.researchgate.net/publication/264808848

Agha N, Syed GK, Mirani DA (2018) Exploring the representation of gender and identity: patriarchal and citizenship perspectives from primary level Sindhi textbooks in Pakistan. Women's Stud Int Forum 66:17–24. https://doi.org/10.1016/j.wsif.2017.11.009

Baker AH, Wardle J (2003) Sex differences in fruit and vegetable intake in older adults. Appetite 40(3):269–275. https://doi.org/10.1016/S0195-6663(03)00014-X

Bárcena A (2020) The emergence of climate change in Latin America and the Caribbean: are we still waiting for catastrophe or moving to action? ECLAC Books, No. 160 (LC/PUB.2019/23-P), Santiago, Economic Commission for Latin America and the Caribbean (ECLAC)

Beardsworth A, Bryman T, Keil J, Goode C, Haslam E, Lancashire (2002) Women, men and food: the significance of gender for nutritional attitudes and choices. Br Food J 104(7):470–491. https://doi.org/10.1108/00070700210418767

Boluda K, Capilla V (2017) Consumer attitudes in the selection of functional foods. Span J Mark – ESIC 21(S1):65–79. https://doi.org/10.1016/j.sjme.2017.05.002

Céspedes RCA (2002) Compatibility between family life and the world of work in Chile: a gender perspective. Final report of the competition: Social fragmentation and political and institutional crisis in Latin America and the Caribbean. CLACSO Regional Fellowship Program. http://bibliotecavirtual.clacso.org.ar/ar/libros/becas/2001/cespedes.pdf

Chávez CJC (2006) Culture of participation and citizenship from a gender perspective. http://www.eumed.net/jirr/1/AMECIDER2006/PARTE%206/292%20Julia%20del%20Carmen%20Chavez%20Carapia.pdf

Colomer RC, Peiró PR (2002) Glass ceilings and slippery ladders? Gender inequalities and strategies for change SESPAS. Gac Sanit 16(4):358–360. https://doi.org/10.1016/S0213-9111(02)71936-5

Cordourier RG (2015) Progress and challenges in gender equality. In: Twenty years after the Beijing platform for action, 1st edn. National Institute for Women [In mujeres], Mexico. https://www.gob.mx/cms/uploads/attachment/file/99985/ct61.pdf

DiClemente DF, Hantula DA (2003) Applied behavioral economics and consumer choice. J Econ Psychol 24(5):589–602. https://doi.org/10.1016/S0167-4870(03)00003-5

Elkhweaky Z, Elbayoumi SI, Manzani YE (2023) Talented women for senior positions across multi-regions: challenges, strategies, and future research agenda. Manag Rev Q Springer. https://doi.org/10.1007/s11301-023-00383-z

FAO (2013). The State of Food and Agriculture: Food Systems for Better Nutrition. Rome. P.C. Young. P.C. (1970). An instrumental variable method for real-time identification of a ruyy process. Automatica 6(2), march 1970, 271-287. https://doi.org/10.1016/0005-1098(70)90098-1

FAO (2013). El estado mundial de la agricultura y la alimentación. Sistemas alimentarios para una mejor nutrición. Roma. https://openknowledge.fao.org/server/api/core/bitstreams/a836bca6-57d3-4893-b82a-be751f1c4767/content

FAO (2017) The future of food and agriculture. Trends and challenges. https://www.fao.org/3/i6881s/i6881s.pdf

FAO (2022) The state of food security and nutrition in the world 2022. https://www.fao.org/3/cc0639en/cc0639en.pdf

Flores VJJ, Gómez CMA, Sánchez PV, Muñoz RM, López GE, Díaz CS (1986) Agroindustry: conceptualization, levels of study, and its importance in agricultural analysis. Agric Geogr J. Regional Studies of Mexican Agriculture. 10–22. https://www.chapingo.mx/revistas/geografia/?section=article&id_articulo=1774&doi=0000

Fraga, C. y Riveiro, M. (2011). La visibilización del género en los estudios de estratificación: el caso de las amas de casa. Nueva época. Temas de mujeres. 7 (7) pp 120-147. https://www.aacademica.org/manuel.riveiro/18.pdf

Freire P (1970) Cultural action and conscientization. Harvard Educ Rev 40(3):452–477. https://doi.org/10.17763/haer.40.3.h76250x720j43175

Girón A (2012) Feminism: breaks and mending in times of crisis. In: Feminism and social change in Latin America and the Caribbean, 1st edn. In: Carosio A (coord) Autonomous City of Buenos Aires, CLACSO

Godoy L, Mauro A (2001) Couple relationships and changes in the labor market: the point of view of men: in Revista de la Academia. No 6, spring 2001, Academia de Humanismo Cristiano, Santiago de Chile

Gupta I, Roy A. (2023). What really empowers women? Taking another look at economic empowerment. Journal of Social and Economic Development. 25:17–31 Springer. https://doi.org/10.1007/s40847-022-00215-y . The online version was published in 2022.

Guzmán V (2002) Gender relations in a global world. Series: Women and Development. United Nations [UN] ECLAC, Chile

Hair JF, Hult GT, Ringle CM, Sarstedt M, Castillo J, Cepeda GA, Roldán JL (2017) Handbook of partial least squares structural equation modeling (PLS-SEM), 2nd edn. Original in English. SAGE Publishing, Spain

Hart SL (1995) A natural-resource-based view of the firm. Acad Manag Rev 20(4). https://doi.org/10.5465/amr.1995.9512280033

Hart SL (2007) Capitalism at the crossroads, 1st edn. Ediciones Deusto, Barcelona, Spain (original in English: Capitalism at the crossroads. Aligning business, earth, and humanity)

Hart SL, Milstein MB (2003) Creating sustainable value. Acad Manag Perspect 17(2). https://doi.org/10.5465/ame.2003.10025194

INEGI (2023a) Monthly indicator of industrial activity. Seasonally adjusted series and trend-cycle (index base 2013=100). www.inegi.org.mx/temas/imai/

INEGI (2023b) National Survey of Household Income and Expenditure (ENIGH) 2022. https://www.inegi.org.mx/contenidos/saladeprensa/boletines/2023/ENIGH2022.pdf

Lamas HM (2005) The importance of domestic work from a regional and gender perspective. Nóesis. J Soc Sci Hum 15(27):101–121. Institute of Social Sciences and Administration: Ciudad Juárez, Mexico. http://www.redalyc.org/articulo.oa?id=85902705

Lancaster K (1966) A new approach to consumer theory. J Polit Econ 74(2):132–157. https://www.journals.uchicago.edu/doi/10.1086/259131

León M (1997) Empowerment in feminist theory and practice. In: Power and empowerment of women. In: León M (ed) Santafé de Bogotá, 1st edn. Tercer Mundo, S.A., Colombia, p 259

Montgomery UW (2011) Behavioral economics and the experimental analysis of consumer behavior. Revista IIPSI 14(1):281–292. Print ISSN: 1560-909X

Ratchford BT (2001) The knowledge economy of the consumer. J Consum Res 27(4):397–411. https://doi.org/10.1086/319617

Ridler CB, Ridler N (2011) A potential conflict between economic and environmental sustainability: a case study MIC 2011: managing sustainability? In: Proceedings of the 12th international conference, Portorož, 23–26 November [Selected Papers], University of Primorska, Faculty of Management Koper, pp 1131–1142. https://www.fm-kp.si/zalozba/ISBN/978-961-266-112-0/papers/MIC4222.pdf

SAUTU, R. (1992): Teoría y medición del estatus ocupacional: escalas ocupacionales objetivas y de prestigio. Cuaderno 10 del Instituto de Investigaciones – Facultad de Ciencias Sociales, UBA. Buenos Aires: IIGG.

Torres, T. F. (1989): The starting point of the modernization of the food supply in Mexico. Momento Económico (45). pp. 29-31. Mexico. https://ru.iiec.unam.mx/1931/

UN Women (2017) SDG 5. Achieve gender equality and empower all women and girls. http://www.unwomen.org/en/news/in-focus/women-and-the-sdgs/sdg-5-gender-equality

UN Women – Schalkwijk (2015) Gender equality. The progress of women in the world. Transforming economies to realize rights. www.unwomwn.org/en

United Nations [UN] (2017) Facts and figures: economic empowerment. The benefits of economic empowerment. UN Women. http://www.unwomen.org/en/what-we-do/economic-empowerment/facts-and-figures

Young PC (1970) An instrumental variable method for real-time identification of a ruyy process. Automatica 6(2):271–287. https://doi.org/10.1016/0005-1098(70)90098-1

Integrated Causal Map of MSMEs in the Creative Sector for Their Promotion in Mexico

10

Citlalli Macías-Barreto, Mario Aguilar-Fernández, Julián Patiño-Ortiz, and Jesús A. Álvarez-Cedillo

Abstract

Creative industries have received special attention in recent years due to their potential contribution to social and economic development in countries with emerging economies. However, these industries tend to be MSMEs with little chance of survival because demand for their products and services is difficult to predict. With a view to achieving target 8.3 of the Sustainable Development Goals, through documentary research with a systems perspective, the variables and causal relationships that influence the growth and development of MSMEs in this sector were identified and mapped. It is concluded that public policies and strategies to promote creative industries in Mexico should be long-term in scope. These policies should be aimed at generating favorable conditions for these industries in: education, research, trade and legal protection of intellectual property, cluster formation, exports, and the creation of regional brands.

Keywords

Cultural and creative industries · Creative sector · Orange economy · MSMEs · Growth and development

10.1 Introduction

Insufficient investment, both public and private, and reduced consumption have led to the continued challenge of creating quality employment in almost all economies. It is estimated that, in 2018, around 30.8% of the population of Latin America and the Caribbean was living below the poverty line; and that in 2022 there were 192 million unemployed people worldwide. Additionally, in many places, having a job does not guarantee quality of life or development (Economic Commission for Latin America and the Caribbean [ECLAC] 2023; United Nations 2023). For this reason, the 2030 Agenda proposes specific measures in which enterprises can collaborate to promote global well-being, thus recognizing the importance of their role in this task (Magos Chávez and Pérez Pineda 2021, p. 20).

One example is Sustainable Development Goal number 8 (SDG 8), which aims to promote economic growth inclusive and sustainable, as

C. Macías-Barreto · J. Patiño-Ortiz
Escuela Superior de Ingeniería Mecánica y Eléctrica Unidad Zacatenco del Instituto Politécnico Nacional, Ciudad de México, Mexico
e-mail: cmaciasb1800@alumno.ipn.mx

J. Patiño-Ortiz
e-mail: jpatinoo@ipn.mx

M. Aguilar-Fernández (✉) · J. A. Álvarez-Cedillo
Unidad Profesional Interdisciplinaria de Ingeniería y Ciencias Sociales y Administrativas del Instituto Politécnico Nacional, Ciudad de México, Mexico
e-mail: maguilarfer@ipn.mx

J. A. Álvarez-Cedillo
e-mail: jaalvarez@ipn.mx

P. Pérez-Hernandez and R. Huerta Cuervo (eds.), *Social Well-Being and the Challenges in the Sustainable Development Goals*, Sustainable Development Goals Series,
https://doi.org/10.1007/978-3-032-14294-8_9

well as employment and decent work for all. Emphasis is placed on target 8.3, which aims to promote public development policies that support productive activity, the creation of decent jobs, entrepreneurship, creativity, and innovation; as well as to encourage the formalization and growth of micro, small, and medium-sized enterprises (MSMEs), including access to financial services (United Nations 2023). In this regard, various sources argue that MSMEs are key to achieving the 2030 Agenda globally, due to their economic significance, their contribution to job creation, and poverty reduction (Magos Chávez and Pérez Pineda 2021, p. 20).

In the case of Mexico, 99.8% of the 6.3 million establishments reported in the 2019 Economic Census are MSMEs (Magos Chávez and Pérez Pineda 2021, p. 22); however, 52% of them tend to cease operations within the first two years (National Institute of Statistics and Geography [INEGI] 2019), a situation that was exacerbated by the health emergency caused by the COVID-19 pandemic, which resulted in the permanent closure of 21.17% of economic units by 2020 (National Institute of Statistics and Geography [INEGI] 2020). In response, some measures were implemented targeting Mexican MSMEs, aiming to improve their liquidity, adaptability, and productive capacity after the pandemic; however, systematic analyses of their exploitation remain pending (Heredia-Zurita and Dini 2021).

In this context, the literature review affirms that Cultural and Creative Industries represent a significant opportunity for social and economic development, especially in countries with emerging economies (Landoni et al. 2020; Rayhan and Subroto 2018; Rodríguez-Gulías et al. 2020). This is because they combine cultural and creative values and meanings with the entire process of producing goods and services—from the conception of an idea to its production, marketing, and distribution. This added value, whether cultural, artistic, or entertainment-based, though intangible, is perceived as greater than strictly material or functional value, and represents a form of potential intellectual capital (Espinal Monsalve 2006; Ghaisani and Subroto 2018). However, these enterprises often remain small with limited chances of survival, due to the high sensitivity of their products to changes in trends and socially constructed perceptions of value, making demand estimation difficult (Crossland and Smith 2002; Espinal Monsalve 2006; Landoni et al. 2020). In other words, demand increases as knowledge of the product deepens, and higher prices are paid based on the anticipated pleasure, which differs from the consumption of non-cultural goods (Espinal Monsalve 2006, p. 78).

The classification of Creative, or Cultural and Creative Industries remains unclear due to their heterogeneity. This sector includes enterprises dedicated to art and the antiques market, advertising, design, fashion, and crafts; as well as music, performing arts, film and video, visual arts and audiovisual media, or new media; and also architecture, publishing, software, and video games (Ghaisani and Subroto 2018; Macías-Barreto and Aguilar Fernández 2021; Vázquez-Álvarez 2022). For the purposes of this document, we will refer to this sector generally as Creative Industries (CIs), recognizing that they are distinguished by their social component, with skills and talents capable of generating wealth and employment through the exploitation of novel and original ideas, and with the potential to translate them into intellectual property (Ahman et al. 2020; Ghaisani and Subroto 2018; Rayhan and Subroto 2018; Vázquez-Álvarez 2022).

To date, statistical analysis of this industry at the international level remains limited, due to the use of different methodologies and criteria for data collection, processing, and presentation, where such data exist (United Nations Conference on Trade and Development [UNCTAD] 2022, p. 26). Nevertheless, it is estimated that these industries have contributed between 0.5 and 7.3% of the Gross Domestic Product (GDP) in countries where data are available (United Nations Conference on Trade and Development [UNCTAD] 2024, p. 1). This is a highly significant figure considering that, for the most part, CIs are composed of independent creators and MSMEs (European Commission 2010, pp. 7–8; United Nations Educational, Scientific and Cultural Organization [UNESCO] 2021, p. 4). Furthermore, in 2020, trade in creative goods between emerging economies accounted for 40.5% of sector exports, surpassing those of developed economies, which export more creative services than goods (UNCTAD 2022, pp. 3–4).

It is worth noting that Latin American countries tend to develop public policies for this sector by prioritizing cultural industries, often neglecting the creative component (Mexican Institute for Competitiveness [IMCO] 2015, p. 7; Piedras Feria 2008, p. 152). In Mexico, recognition of the potential of CIs has been slower and more recent compared to other Latin American countries (Castañeda Rivera and Garduño Bello 2017, p. 122). As in other countries, the economic activities of the creative sector have been identified through the implementation of the Satellite Account System for Culture (CSC) by the National Institute of Statistics and Geography and the National Council for Culture and the Arts, based on frameworks from the United Nations Educational, Scientific and Cultural Organization (UNESCO), the Andrés Bello Agreement, and the World Intellectual Property Organization (WIPO). This tool measures 10 economic sectors: Heritage; Performing arts; Books, printing, and press; Trade in cultural products; Music and concerts; Visual arts and photography; Public management in cultural activities; Audiovisual media; Design and creative services; and Crafts and traditional toys (Economic Commission for Latin America and the Caribbean [ECLAC] 2014, p. 87; Sánchez-Jofras and Kuri-Alonso 2020, pp. 70–71).

Based on 2018 data, the CSC estimates that in 2022, the CIs sector in Mexico accounted for 2.9% of GDP, equivalent to 815,902 million pesos, with a total of 1,494,745 jobs; representing 3.6% of the national total. The four activities with the highest GDP contribution for the same year were: Handicrafts, with 19.3%; Digital content and internet, with 18.6%; Audiovisual media, with 18%; and Design and creative services, with 12.8%, respectively; 31.3% is distributed among the other six activities (National Institute of Statistics and Geography [INEGI] 2023). Following the COVID-19 pandemic, it is still necessary to update these statistics, such as the percentage of creative establishments, which in 2018 was 2% of the total 5 million economic units in Mexico (British Council 2019, p. 9), mainly located in Mexico City and the State of Mexico, as well as in specific regional clusters, such as the software cluster in Nuevo León and Jalisco, and the handicrafts cluster in Puebla and Guerrero (Ministry of Culture 2019).

It is worth noting that Mexico is the country with the most archaeological sites in Latin America, with a total of 162 out of the 308 recorded in the region (ECLAC 2014, p. 232). There is also a Mexican Network of Creative Cities, created in 2020 to strengthen these hubs through technical support, guidance, and collaboration. It is composed of: Ensenada and Mérida, gastronomy cities; Guadalajara, city of digital art; Morelia and Xalapa, music cities; Puebla, Mexico City, and Querétaro, design cities; and San Cristóbal de las Casas, city of handicrafts and folk arts (United Nations Educational Scientific and Cultural Organization 2023). This demonstrates recent efforts by the Mexican government to mobilize the creative community in designing growth strategies, as well as promoting youth entrepreneurship and exchange programs, training cultural leaders and public officials in policy formulation for the sector in collaboration with the United Kingdom (British Council 2019).

Another notable initiative is *México Creativo*, whose purpose is to map the creative economy in the country. This involves the creation of a *transmedia* map developed through collaboration among various entities, led by the Digital Culture Center of the Ministry of Culture of the Government of Mexico, with the participation of the British Council, the Inter-American Development Bank, the Universidad Autónoma Metropolitana-Xochimilco, and the NESTA Foundation (Ministry of Culture 2019). Key findings from this work include: 54.3% of creatives work independently; only 21% are employed by a company or organization; and 24.7% combine both modalities. Regarding the size of creative enterprises: 55.7% are microenterprises; 28% are small businesses; and only 16.3% are medium and large enterprises. The profile of entrepreneurs in this sector shows that: 89% are between 25 and 60 years old;

the majority have a professional degree, 56.9%; followed by postgraduate studies, 24.4%; and only 21.7% have an education level below upper secondary. These aspects, together with income data, indicate that these entrepreneurs prioritize their passion over economic benefits, as 76.8% earn less than 12,000 Mexican pesos per month (México Creativo 2021).

In several developing countries, solid and coherent strategies to promote the CIs and leverage them as drivers of economic development have yet to be established. This became evident during the COVID-19 pandemic, as creative sector entrepreneurs were affected by prolonged lockdowns. This highlights that creative goods and services are not considered essential and their consumption largely depends on human gatherings, which poses a challenge for these industries—not only in terms of financing, but also in the sustainability of their operations, which rely on the constant generation of innovative solutions (Rozentale and van Baalen 2021). Despite efforts to quantify CIs activity in Mexico, the available statistical information is not based on recent censuses that would allow for a better characterization of the sector. This hinders the design of coherent public policies and robust government strategies to drive growth and economic development through the promotion of CIs, which, according to the Mexican Institute for Competitiveness (IMCO 2015, p. 3), remained below the national average in 2015.

Due to this issue, the central objective of this research, aligned with SDG 8, was to identify in the specialized literature the variables and relationships that influence the growth and development of CIs, with the aim of generating an integrated causal map that will be useful for the formulation of public policies and coherent strategies to promote creative sector MSMEs in Mexico, using a systems perspective.

10.2 General Guidelines for Causal Mapping

There is a large body of literature on the use of causal mapping across a variety of disciplines since the 1970s, including its use as a decision-making support tool (Powell et al. 2023), as it is a technique for illustrating the existence of feedback within and between subsystems, visualizing mutually dependent variables as a system of relationships, and thus showing various concepts connected by their causalities according to our sensemaking processes (Castaño-Barreto et al. 2020; Woodside 2016).

Powell et al. (2023) state that the definition of a causal map does not vary among most authors, understanding it as: a diagram or graphical structure in which factors (or nodes) are linked by arrows with a direction, representing the sense of causality between them, using positive and negative signs (+) or negative (−). Such visualization helps to capture the complexity of the feedback relationships present (Woodside 2016). A variable is considered dependent if it only has arrows pointing toward it; independent if it only has arrows pointing away from it; and interdependent if it has arrows in both directions (Weick 1979, p. 72, cited in Woodside 2016). Likewise, a causal or feedback loop R is a relationship or chain of relationships that begins and ends with the same variable. A loop may be more important than others, either because it has a greater number of inputs and outputs, or because it contains a larger number of elements (Woodside 2016).

The constituent parts of causal maps are based on beliefs or evidence whose veracity the mapper does not examine in depth, and therefore tend to be rather unsophisticated; since there is no attempt to encode the rich variety of causal statements found in ordinary language, this allows for the handling of a large amount of evidence. It is based solely on the assertion that X influences Y, either positively or negatively. In this sense, causal maps are useful not only as models of beliefs or facts, but also as repositories of evidence from different sources, to organize, combine, present, and draw inferences from a large number of relatively unstructured causal statements.

Given the large number of variables that may be involved, it is important to avoid ending up with a tangled web of meaningless networks; therefore, it is useful to derive several smaller maps from the global map based on the characteristics of the data, although this introduces the possibility of selection bias (Powell et al. 2023). To overcome this selection bias, the viable systems model was chosen to encompass the variables identified after the theoretical analysis of specialized literature, since the purpose of mapping is to achieve the integration of the involved parts in order to obtain a comprehensive interpretation of the evidence.

10.3 Organizational Cybernetics Approach: Viable Systems Model

When addressing an object of study from a systems perspective, it is necessary to consider it as being in constant interaction with the subsystems that comprise it, and simultaneously, as part of broader systems to which it belongs. As a result of applying an integrative approach in research, one often ends up with a large amount of data, variables, and interactions that are difficult to interpret as a whole. Therefore, it is useful to employ a reference and analysis model that allows for the identification of the structural level at which the system under study is located, and how it relates to other systems.

Based on the approach of complex and adaptive systems, both in terms of structure and function, and with the aim of facilitating data analysis, Organizational Cybernetics (OC) was selected; in particular, the Viable Systems Model (VSM) proposed by Stafford Beer (1985), as it provides a functional recursive structure applicable to the analysis of all types of organizations. According to Beer (1985, p. ix), cybernetics addresses the science of efficient organization, examining the laws and principles of communication and control applicable to animate and inanimate systems, whether technical, social, or sociotechnical.

The VSM is configured as a conceptual tool derived from the principles and axioms that govern a viable entity in terms of its dynamic structures and the interconnection of its components for adaptation. It is used as a reference in organizational analysis, with the purpose of detecting deficiencies, inconsistencies, or absences in its structure and connections (Reynolds and Holwell 2010). In this sense, the VSM constitutes a conceptual framework that enables the modeling of conditions for viability in complex systems; it is structured in five levels or systems that operate in an integrated manner within their specific environment.

Accordingly, this approach is used as a reference to analyze the functioning of organizations, detecting weaknesses, misalignments, or deficiencies, based on the premise that organizations resemble living systems that must meet certain viability criteria to maintain both internal and external equilibrium. It is common to understand a viable company as one whose activity is economically profitable; while this is undeniably of great interest in business, it does not mean it is the ultimate goal. As presented here, a company is viable if it operates as a complex system, capable of adapting to constant changes in its environment, since its survival depends on its ability to maintain equilibrium in relation to various environmental conditions (Beer 1985). To achieve this, it must be able to employ various mechanisms to reduce environmental turbulence by increasing its own internal variety (Mugurusi and de Boer 2019).

According to Beer (1985), a system must be composed of a minimum set of crucial subsystems, without which the system's viability would be compromised. Figure 10.1 graphically represents the VSM as a series of critical, interconnected components, arranged recursively in five levels or subsystems that together form a larger system or whole. The necessary interconnections for its viability are also represented, as well as the specific external environment of the system under analysis. To facilitate interpretation, Table 10.1 presents a summary of the function of each subsystem in the model. Readers interested in further details are encouraged to consult the work of Reynolds and Holwell (2010).

10.4 Methodological Framework

This section provides a brief explanation of the methodological strategy chosen to answer the research question: What are the variables and rela-

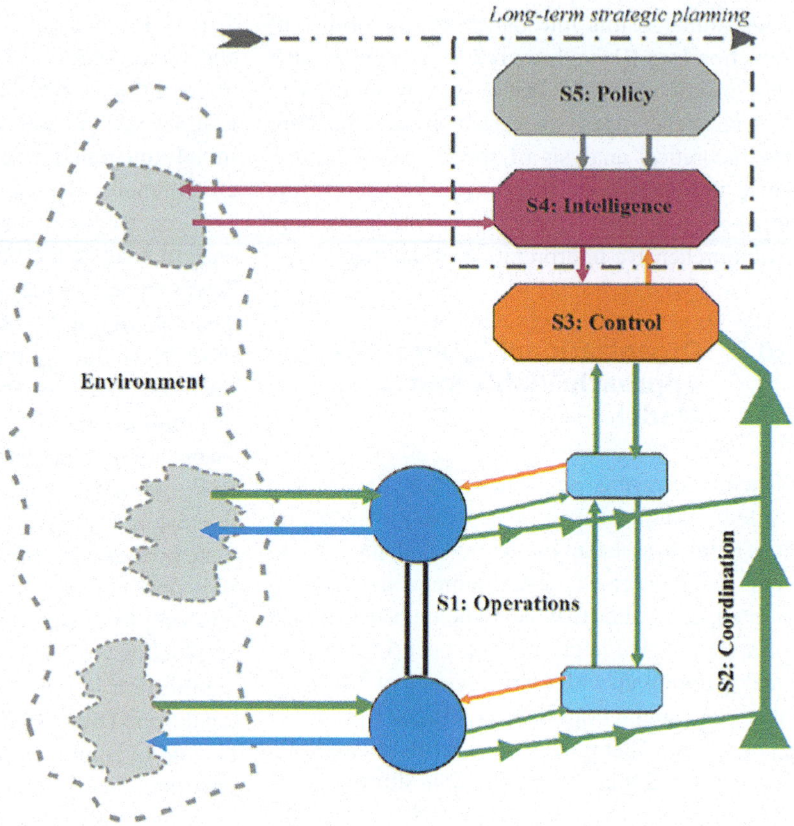

Fig. 10.1 Representation of the structure and components of a viable system. *Source* Compiled by the author based on Beer (1985)

Table 10.1 Summary of the functions of the subsystems in a viable systems model

Subsystem	Function
System 1 (S1) Operations	This subsystem encompasses the production activities and delivery of goods/services that provide value to the external environment. These activities are the driving force of the organization, making their efficiency and effectiveness vital for the system's viability. It is represented by a set of blue circles and rectangles
System 2 (S2) Coordination	This subsystem is responsible for allocating resources and coordinating activities among departments to contain the causes of potential conflicts. It is represented by a series of green triangles on the right side of the diagram
System 3 (S3) Control	This subsystem generates performance indicators to measure and provide feedback to the organization, and takes corrective actions to ensure operational performance. It is represented by an orange rectangle
System 4 (S4) Intelligence	This subsystem includes strategic planning, innovation, and continuous improvement. It is future-oriented and focused on understanding the organization's environment and seeking opportunities for growth. It is represented by a magenta rectangle
System 5 (S5) Policy	This subsystem is responsible for preserving the organization's identity through its management function. Here, policy, vision, mission, and values are established, serving as the basis for aligning the other subsystems. It also ensures balance among the subsystems, especially between operations and intelligence. It is represented by a gray rectangle
Environment	The environment is represented outside the system, conventionally as an undefined droplet shape located on the left side of the diagram. It encompasses all variables, actors, and factors external to the organization, over which it has no control, but which do affect its equilibrium

Source Own elaboration based on Beer (1985)

Table 10.2 General aspects of the research

Type and scope	Description
Research line in the systems engineering doctorate	Complex systems
By object of study	Basic
By depth	Descriptive
By sources of information	Documentary
By approach to the object	Qualitative

Source Own elaboration based on Pacheco and Cruz (2006)

tionships that influence growth and development for the formulation of coherent public policies and strategies to promote micro-, small-, and medium-sized enterprises (MSMEs) in the creative sector in Mexico, from a systems perspective? The general aspects are summarized in Table 10.2.

1. **Data Collection**

A traditional literature search with a systems approach was conducted in February 2023, using the Web of Science and SCOPUS databases to locate articles and literature reviews published from 2019 to 2023, following the three stages detailed in Table 10.3.

2. **Data Analysis**

The 36 selected documents were thoroughly reviewed to identify causal relationships, which were recorded in a spreadsheet. Using the functions of the VSM subsystems described in Table 10.1, the identified relationships were processed using the *axial coding* method (Hernández Sampieri et al. 2010, p. 494). In this way, each causal relationship identified in the literature was placed in a subsystem of the VSM. Subsequently, only the relationships that occurred between the CIs and their environment were retained, resulting in a total of 24 documents.

3. **On Data Processing and Presentation**

The causal relationships identified in the literature were linked using the deductive method with a word processor. This phase was carried out simultaneously with the support of the vector program CorelDRAW 2020 (Corel Corporation 2020). This method enabled the creation of simple links at an initial level, gradually increasing the complexity of the map as more links were integrated, both graphically and in the word processor. In this way, constructs or categories emerged, composed of chains of causality or feedback with lower-level variables. A total of seven chains, referred to as feedback loops (R), were created and served as the inputs for the map. For graphical organization, the criteria of the VSM structure described in

Table 10.3 Steps for traditional literature search

Step 1. Conceptual boundaries for document search in WOS and SCOPUS	Step 2. Definition of criteria for document search in WOS and SCOPUS	Step 3. Definition of exclusion criteria for retrieved documents
1. First-level key concepts: "system dynamics", modeling, "causal maps", "viable systems model", "sustainable development goals" 2. Second-level key concepts: "creative industries", "cultural industries", enterprise*, SMEs. A total of 52 searches were designed 3. Search period: the last five years from the date of the most recent publication. A total of 1,948 documents were obtained	4. Documents: Articles and review articles. All documents were included when results were fewer than 100 5. Research areas: Business, engineering, development studies, management, business economics, and public administration As a result, 51 documents were retained	6. Due to duplication 7. Not related to research on CIs As a result, 36 documents were obtained 8. Did not present causal relationships between the SME and its environment Finally, a total of 24 documents were retained

Source Own elaboration based on Pret and Cogan (2019)

Table 10.1 were used; thus, secondary maps were derived, each identified with a color code.

10.5 Results

After identifying and coding the relevant variables for this research, they were grouped into seven spheres based on the VSM criteria, which form the structural basis of the Integrated Causal Map of Creative Sector MSMEs (see Fig. 10.2), composed of two main blocks: the *micro environment*, and the *environment*. To interpret the map, it is recommended to consult the color code in Fig. 10.3 to facilitate the identification of the VSM spheres and the variables grouped within them.

The *micro environment* consists of the five VSM subsystems, namely: policy, intelligence, control, coordination, and operations. Feedback loops R1, R2, R3, R4, and R5 are located at this level, as they begin and end with variables grouped within one of the spheres of the MSME subsystems.

The *environment* is divided into two levels: the *meso environment* and the *macro environment*. When a variable or actor is placed at the meso level, it refers to the immediate environment of the MSME; whereas, at the macro level, it refers to the regional or national environment. Feedback loops R7 and R6 are located at this level, as they begin and end with variables grouped within the MSME's environment.

Below is a brief interpretation of the five loops located in the micro environment.

Loop R1 illustrates the process of value creation for the environment, which begins with the creation of goods and services with a distinctive character (He and Fu 2014) and ends with the receipt of revenue and profits. In this process, CIs

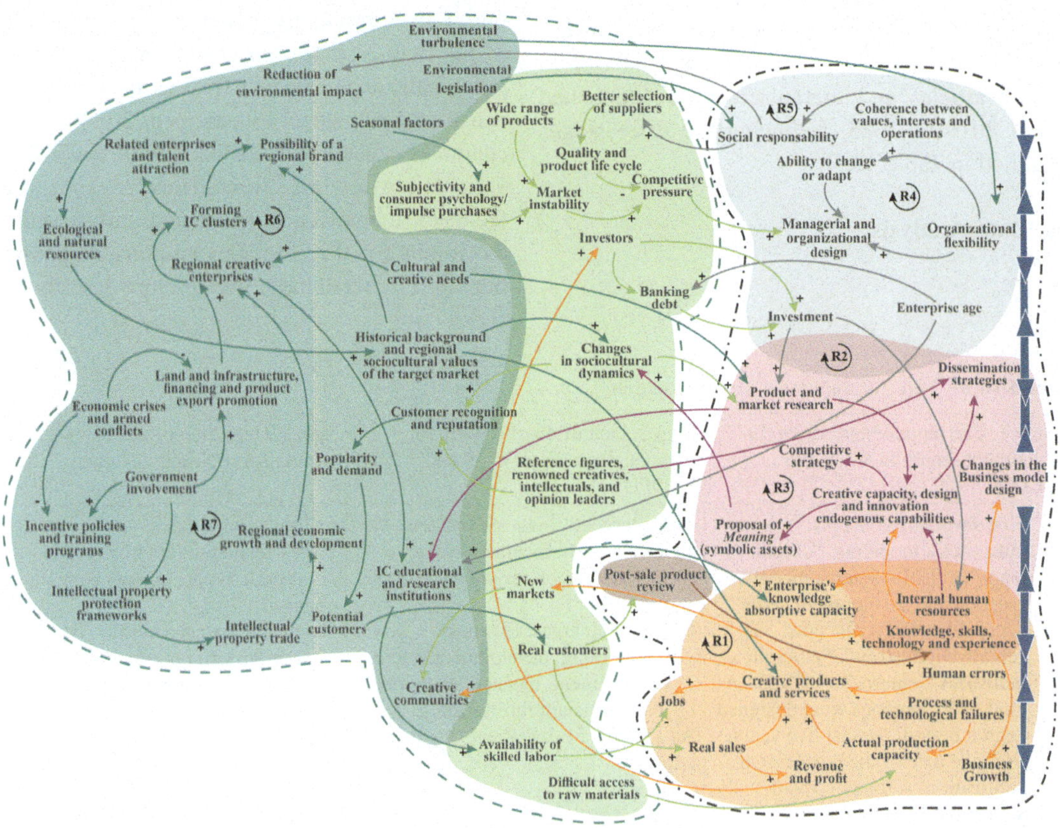

Fig. 10.2 Integrated Causal Map of Creative Sector MSMEs. *Source* Own elaboration based on the theoretical analysis of the literature review

Fig. 10.3 Color key for interpreting the Integrated Causal Map of Creative Sector MSMEs. *Source* Own elaboration based on Fig. 10.2

also generate jobs and contribute to the creation of new markets and creative communities (Ahman et al. 2020). The creative and innovative capacity, as well as the independence from external sources of knowledge, of enterprises increases with their age and with the availability of internal human resources. In this sense, the younger the company, the more guidance it will require, whether from schools or research institutions, intellectual property markets, or licensing (Rodríguez-Gulías et al. 2020; Yang and Qi 2022). If the MSME aims to consolidate sales and the resulting revenue to attract potential investors and reduce its need for debt, it must consider its actual production capacity. Some factors that affect production in any company include technological and process failures, human errors, access to raw materials, inflexible payment terms with suppliers, among others (Abdel-Latif et al. 2022; Ghaisani and Subroto 2018; Liu et al. 2022).

Loop R2 illustrates the process of business growth, which starts with investment and ends with the design of the business model. Dependence on external sources of knowledge can be overcome by improving the company's absorptive capacity and investing in product research through debt (Rodríguez-Gulías et al. 2020). The available human capital in CIs is more relevant than their surrounding physical infrastructure, as increased knowledge, skills, experience, and technological capabilities enhance their innovation capacity for value creation, which in turn drives business growth and changes in the business model design (Ahman et al. 2020; Landoni et al. 2020; Rodríguez-Gulías et al. 2020).

Investment in human resources influences the company's creative capacity and its dissemination and competitive strategies (Mwasubila et al. 2022; Sopelana et al. 2021; Yasin et al. 2019), which supports the generation of creative products and services, increasing the number of potential customers and their demand; this, in turn, leads to an increase in the number of customers and actual sales (Ghaisani and Subroto 2018; Mwasubila et al. 2022). It is worth noting that dissemination strategies for creative products are often exhibitions and fairs (Sopelana et al. 2021); however, dissemination through civil sociability is a promotional method that fosters a culture of use for these products, leveraging socially recognized reference figures such as intellectuals, other renowned creatives, and opinion leaders (Ahman et al. 2020).

Loop R3 presents the process of meaning generation, which begins with understanding sociocultural dynamics and ends with influencing them. In addition to traditional demand drivers, CIs possess a distinctive character that Paris and Ben (2019) refer to as *meaning*, and that Landoni et al. (2020) call *symbolic assets*. In this loop, it can be seen that these products and services are distinguished by this meaning, which consists of translating social, cultural, or political values during the endogenous design phase and through endogenous innovation capabilities. These meanings make the products more attractive to the target market and increase their demand, thereby socializing the cultural apparatus itself (Pitts 2022; Wilson 2020). The meanings are enriched by understanding the historical aspects, values, and sociocultural dynamics observed during the target market research phase and in the post-sale product review (Ghaisani and Subroto 2018). The resulting proposals for meaning generate new uses of language and changes in the sociocultural regime (Paris and Ben 2019; Rodríguez-Gulías et al. 2020). These changes reflect customer recognition, which generates reputation and boosts demand (Landoni et al. 2020).

Loop R4 shows the process of *organizational flexibility* that begins with managerial and organizational design and ends with its review. Organizational flexibility represents the organization's ability to change or adapt in response to *environmental turbulence*. When this occurs, managerial and organizational design tasks are stimulated (Sopelana et al. 2021). The response capacity operates under competitive pressure and is directly related to time and budget, with uncertain outcomes (Pitts 2022) due to market instability caused by seasonal factors, a wide range of products, subjective evaluation criteria, and consumer psychology that drives impulse purchases (Wilson 2020).

Loop R5 represents the process of identity creation for the company, which begins with social responsibility and ends with the generation of creative goods and services. *Social responsibility* refers to the alignment between a company's operations, its values, and the interests of all stakeholders.

This entails complying with the environmental requirements legally established by the government, which includes optimizing the selection of suppliers to ensure the expected quality, as well as assessing environmental impact, product life cycle, and implementing reasonable scheduled or preventive maintenance (Franco 2019; Liu et al. 2022; Wasserbaur and Sakao 2020).

It is important to note that the variables and relationships of control and coordination functions have received little attention in the reviewed literature, and thus their presence in the map is scarce. Having presented the microenvironment loops, a brief interpretation of the two loops located in the MiPyme's external environment is provided below.

Loop R6 represents the process of forming CIs *clusters*, which begins with the identification of public cultural and creative needs and culminates in the creation of regional brands. According to He and Fu (2014), as the number of creative enterprises in a region increases, so does the number of educational and research institutions in CIs, consequently leading to a rise in qualified human resources. As Buitrago Restrepo and Duque Márquez (2013, p. 196) state, "empires of the mind are not built with buildings but with communities." In this sense, the proliferation of creative enterprises can be leveraged for the formation of *clusters*. He and Fu (2014) argue that this strengthens the attraction of related enterprises and more talent to the region, due to the intensive knowledge demand required for sustainable development. This, in turn, increases the likelihood of establishing a regional brand based on historical background and regional sociocultural values linked to the ecological environment and its natural resources.

Loop R7 represents the process of reducing turbulence, which starts with government support and ends with regional growth and development. Government involvement at all levels is key to the clustering and development of these enterprises (Ahman et al. 2020), by providing land and necessary infrastructure, as well as offering incentive policies, training programs, financing, support for product export, and the trade of intellectual property owned by the enterprises, along with the establishment of appropriate legal protection frameworks (Abdel-Latif et al. 2022; Yang and Qi 2022). As can be seen, government and civil society participation, cooperation among enterprises with social and environmental responsibility, and the promotion of educational and research institutions are all relevant for regional economic growth and development (Ahman et al. 2020; Carayannis et al. 2012; Rodríguez-Gulías et al. 2020; Yang and Qi 2022). This contribution could be measured by considering the benefits to intellectual property rights (Yang and Qi 2022), since operating within a governmental framework makes them easily quantifiable. It should be noted that, at this level, economic crises and armed conflicts are also identified (Abdel-Latif et al. 2022), situations that logically reduce the government's support capacity. Government intervention at all levels (the greater the synergy among them, the greater the development of enterprises) and its regulation are crucial, as their actions help reduce environmental turbulence, allowing MSMEs to consolidate their investments for growth and development. Nevertheless, it is necessary to recognize that the most valuable asset of these enterprises is the profile of their human resources, and their most important investment is in their capabilities and talent (Buitrago Restrepo and Duque Márquez 2013, pp. 174–175).

10.6 Conclusions

In pursuit of SDG 8, which aims to promote inclusive and sustainable economic growth, employment, and decent work, specifically in relation to target 8.3, which calls for the promotion of public policies to support productive activities, job creation, entrepreneurship, creativity, and innovation, as well as to encourage the formalization and growth of micro-, small-, and medium-sized enterprises. In this chapter, a systems approach was used to identify, through theoretical analysis of the specialized literature, the variables and relationships that influence the growth and development of CIs. As a result, the integrated causal map is presented, contributing to the formulation of coherent public policies and strategies to foster creative sector MSMEs in Mexico. It is concluded that these enterprises are particularly sensitive to environmental turbulence. In this regard, to foster their growth and development, a long-term approach is recommended; the following recommendations are suggested.

Investment in schools and public research institutions in CIs is a factor whose relevance was foreseeable; the availability of human resources with creative profiles is key to the conception and consolidation of an enterprise, even more so than infrastructure. For this reason, the planning and location of training and research institutions should consider the highest concentration of creative commercial activities. This implies conducting a census of the creative economy in Mexico, following a comprehensive categorization of both creative and cultural industries.

To drive economic growth by fostering CIs, the formation of creative *clusters* is necessary to increase their opportunities through cooperation and resource pooling, enabling them to optimize their productive and creative capacity for growth. In this regard, given the size of the national territory and the distribution of its poorest population, the concentration of creative enterprises in Mexico City, the State of Mexico, and the *clusters* of Nuevo León and Jalisco does not directly contribute to the growth and development of the country's most marginalized regions. Therefore, promoting the artisanal *clusters* of Puebla and Guerrero is a positive step. However, due to their recent establishment, there is still no information available to assess their results and determine their economic impact, although it is already possible to anticipate that their number remains unrepresentative of CIs and the national territory.

It is advisable to continue promoting these clusters throughout the rest of the country, which involves encouraging the export of creative products and services, but above all, fostering a commercial market for the intellectual property generated by enterprises. This is essential because, for a creative activity with economic benefits to become part of the CIs, the creation of intellectual property is necessary. Therefore, it is crucial to review and improve the legal frameworks for the protection of all types of creative products and services in Mexico, with the aim of ensuring investment in research and development, both technological and creative or cultural. In this regard, one alternative is the creation of *creative hubs* to develop, produce, and commercialize goods and services within the creative economy, by concentrating interactions among capital, talent, and technology, regardless of geographic proximity.

On the other hand, it is recommended to promote collaboration among enterprises, civil society, and research institutions to generate annual reports on regional consumption trends, in order to identify needs that can be addressed by the creative industries. Likewise, the creation and consolidation of *regional brands* should be encouraged with the active participation of local producers, since this approach transcends the creative limitations of *designation of origin.* The regional brand encompasses a wide range of products whose commercial identity is associated with a specific region, lifestyle, sociocultural values, and geographic characteristics, without limiting the commercial identity to the locality's name or to a single product whose features must be intrinsic to its geographic origin. The creation of regional brands is a dissemination strategy with greater reach than exhibitions and fairs, as it is a concept that accompanies products without the need for direct sales efforts. In this sense, the socialization of the regional brand is more effective in fostering a culture of use for these products by identifying socially recognized reference figures. Promoting the creation of regional brands requires engaging in empathetic dialogue with creative producers and providing incentives to ensure the initiative succeeds in harmoniously unifying different perspectives, values, and identities. This is essential for producers to feel that the brand represents them coherently.

It is recommended to consider establishing preferential environmental legal frameworks for the creative industries, given that they are not typically resource-intensive. In light of sociocultural trends, a brand with a social responsibility certification can increase its appeal, thereby reducing competitive pressure. This requires a clear definition and characterization of beneficiary enterprises to prevent policy benefits from enabling the evasion of corporate social responsibility, as well as updated measurements to estimate the environmental impact of each company.

Finally, according to the literature reviewed, collaboration among government, civil society, socially responsible enterprises, and education and research in the creative industries is key to the growth and development of the sector. However, this mapping has not considered the influence of various interests—local, national, and international—that affect the process of public policy design. In pursuit of SDG 8, when formulating policies to promote the growth and development of creative sector MSMEs in Mexico, it is essential to acknowledge the lack of statistical information on the sector by involving more analysts and disseminating among creative industry stakeholders the advantages of making decisions based on these quantitative tools. Likewise, it is important to avoid contradictions such as the one already mentioned regarding the location of *clusters* and the legal gap in intellectual property protection, as well as the negotiated conditions in current international trade agreements, whose clauses have committed the development of key sectors such as technology and film.

The results presented here represent an initial approach to understanding the complexity in which the creative industries operate, by integrating the main variables and causal relationships identified in previous mappings of creative industry growth and development (Ghaisani and Subroto 2018; Sopelana et al. 2021; Yang and Qi 2022), as well as regional economic growth and development related to them (Abdel-Latif et al. 2022; He and Fu 2014; Wilson 2020). Although the wide variety of subsectors, the heterogeneity of the creative industries, and the limited statistical information make generalizations difficult, it is up to researchers to identify similarities, trends, and consistencies among creative industry subsectors that are useful for analyzing the sector as a whole. For this purpose, this study used the MSV structure to encompass the relevant variables and relationships identified, which facilitated the integration of information, overcoming the specificity of each case and potential selection bias. This integrated causal map is a preliminary step toward the development of quantitative system dynamics models that allow for the simulation of hypothetical scenarios to less uncertainty address the complex relationships in which creative sector MSMEs operate, since causal maps alone cannot achieve this.

References

Abdel-Latif A, Saad-Eldien A, Marzouk M (2022) System dynamics applications in crisis management: a literature review. J Simul 00(00):1–18. https://doi.org/10.1080/17477778.2022.2088306

Ahman E, Machmud A, Sumawidjadja RN, Herlinawati E (2020) Model of the competitive advantage of creative industries in Indonesia based on helix quadruple, creativity and innovation capability. Int J Innov Creativ Change 11(8):516–532

Beer S (1985) Diagnosing the system for organizations. Oxford University Press

British Council (2019) Skills for the creative economy: needs, training, and learning styles in Mexico. https://bit.ly/3S6cyGg

Buitrago Restrepo F, Duque Márquez I (2013) The orange economy: an infinite opportunity. In: Inter-American Development Bank, vol 1. https://bit.ly/4bqEYD2

Carayannis EG, Barth TD, Campbell DF (2012) The Quintuple Helix innovation model: global warming as a challenge and driver for innovation. J Innov Entrep 1(1):2. https://doi.org/10.1186/2192-5372-1-2

Castañeda Rivera E, Garduño Bello B (2017) Map of the creative industries in Mexico. Projection for CENTRO. In: Creative economy (Issue 7). https://doi.org/10.46840/ec.2017.07.05

Castaño-Barreto AC, Jaramillo-Cruz CA, Molina Benavides RA, Atzori AS (2020) Scenarios of sustainable fishing in the Zapatosa marsh (Colombia) simulated with a system dynamics model. Sustainability (Switzerland) 12(8). https://doi.org/10.3390/SU12083458

Corel Corporation (2020) CorelDRAW graphics suite (No. 2020). [Graphics software]. https://bit.ly/4bBNteb

Crossland P, Smith FL (2002) Value creation in fine arts: A system dynamics model of inverse demand and information cascades. Strateg Manag J 23(5):417–434. https://doi.org/10.1002/smj.231

Economic Commission for Latin America and the Caribbean (2014) Culture and economic development in Ibero-America. https://bit.ly/4bBCD8j

Economic Commission for Latin America and the Caribbean (2023) Goal 8. Decent work and economic growth. 2030 Agenda in Latin America and the Caribbean regional knowledge platform. https://bit.ly/3JVfGRF

Espinal Monsalve NE (2006) Cultural economics. In: UN electronic journals essays on economics. https://repositorio.unal.edu.co/handle/unal/36923

European Commission (2010) Green paper: unlocking the potential of cultural and creative industries. In: Green paper on cultural and creative industries. https://bit.ly/3zU41gO

Franco MA (2019) A system dynamics approach to product design and business model strategies for the circular economy. J Clean Prod 241:118327. https://doi.org/10.1016/j.jclepro.2019.118327

Ghaisani AD, Subroto A (2018) The model of craft subsector's development within Indonesia's creative industry: a system dynamics approach. In Soliman KS (ed) 32nd Conference of the international business information management association (IBIMA), pp 2622–2636. https://bit.ly/3UvpbLM

He Y, Fu L (2014) Simulation model of cultural industry agglomeration based on SD-system dynamics theory. Key Eng Mater 584:312–317. https://doi.org/10.4028/www.scientific.net/KEM.584.312

Heredia-Zurita A, Dini M (2021) Support policies for MSMEs in the COVID-19 context in Mexico. In: Economic Commission for Latin America and the Caribbean (ed) Analysis of support policies for SMEs to address the COVID-19 pandemic in Latin America (Documents). United Nations, pp 87–93. https://bit.ly/4by7dzp

Hernández Sampieri R, Fernández Collado C, del Baptista Lucio MP (2010) Research methodology, 5th edn. McGraw Hill

Landoni P, Dell'era C, Frattini F, Messeni Petruzzelli A, Verganti R, Manelli L (2020) Business model innovation in cultural and creative industries: Insights from three leading mobile gaming firms. Technovation 92–93(August 2019):102084. https://doi.org/10.1016/j.technovation.2019.102084

Liu P, Hendalianpour A, Hafshejani MF, Yaghoobi F, Feylizadeh M (2022) System dynamics model: developing model for supplier selection with a focus on CSR criteria. Complex Intell Syst. https://doi.org/10.1007/s40747-022-00788-5

Macías-Barreto C, Aguilar Fernández M (2021) Viable business model for creative enterprises for small and medium-sized craft businesses. Nexo Sci J 34(04):1129–1139. https://doi.org/10.5377/nexo.v34i04.12637

Magos Chávez R, Pérez Pineda JA (2021) Key factors for collaborating with Mexican micro, small, and medium-sized enterprises to achieve the 2030 Agenda through international cooperation: the Spanish case "SMEs and SDGs." Revista Int Cooper Dev 8(2):19–32. https://doi.org/10.21500/23825014.5560

Mexican Institute for Competitiveness (2015) Creative industries & protected works. https://bit.ly/47zpj1G

México Creativo (2021) Perception of the impact of COVID-19 on the cultural and creative economy of Mexico. 2021 Survey. https://bit.ly/3U511tb

Ministry of Culture (2019) The transmedia map is presented, a detailed atlas of Mexican creative industries. https://bit.ly/3QDl4wo

Mugurusi G, de Boer L (2019) Are you receiving me? A viable system model (VSM) analysis of purchasing coordination in a firm engaged in offshoring of manufacturing activities. Syst Pract Action Res 32(3):239–272. https://doi.org/10.1007/s11213-019-09483-y

Mwasubila IJ, Taifa IWR, Kundi BAT (2022) An analytical study on establishing strategies for improving the productivity of the spinning industries. Int J Ind Syst Eng 40(1):1–28. https://doi.org/10.1504/IJISE.2022.120804

National Institute of Statistics and Geography (2019) Business demography 1989–2019. Mortality. https://bit.ly/3KHz6cq
National Institute of Statistics and Geography (2020) Results of the second edition of ecovid-ie and the 2020 business demography study. In: Press Release No. 617/20 (Issue 00). https://bit.ly/40dxeOx
National Institute of Statistics and Geography (2023) Satellite account of culture of Mexico, 2022. https://bit.ly/3QHnPN4
Pacheco A, Cruz MC (2006) Critical research methodology. In: Grupo Patria Cultural (ed) Logic, procedure, and techniques, 1st edn
Paris T, Ben Mahmoud-Jouini S (2019) The process of creation in creative industries. Creativ Innov Manag 28(3):403–419. https://doi.org/10.1111/caim.12332
Piedras Feria E (2008) Mexico: cultural technology for comprehensive development. In: Reis ACF (ed) Creative economy as a development strategy: a view from developing countries. Itaú Cultural, pp 150–167. https://bit.ly/3QH9791
Pitts FH (2022) Measuring and managing creative labour: value struggles and billable hours in the creative industries. Organization 29(6):1081–1098. https://doi.org/10.1177/1350508420968187
Powell S, Copestake J, Remnant F (2023) Causal mapping for evaluators. Evaluation. https://doi.org/10.1177/13563890231196601
Pret T, Cogan A (2019) Artisan entrepreneurship: a systematic literature review and research agenda. Int J Entrepren Behav Res 25(4):592–614. https://doi.org/10.1108/IJEBR-03-2018-0178
Rayhan MN, Subroto A (2018) The development model of the creative industry culinary subsector in Indonesia: system dynamics approach. In: Soliman KS (ed) 32nd Conference of the International-Business-Information-Management-Association (IBIMA), pp 2637–2649. https://bit.ly/3o4wNc1
Reynolds M, Holwell S (2010) Systems approaches to managing change: a practical guide. In: Reynolds M, Holwell S (eds) The Open University, 1st edn. Springer, London. https://doi.org/10.1007/978-1-84882-809-4
Rodríguez-Gulías MJ, Fernández-López S, Rodeiro-Pazos D (2020) Innovation in cultural and creative industries firms with an academic origin (CCI-USOs): the role of regional context. Technovation 92–93(January 2017):1–10. https://doi.org/10.1016/j.technovation.2018.06.007
Rozentale I, van Baalen PJ (2021) Crafting business models for conflicting goals: lessons from creative service firms. Long Range Plan 54(4):102092. https://doi.org/10.1016/j.lrp.2021.102092
Sánchez-Jofras JF, Kuri-Alonso I (2020) Employment distribution in the cultural and creative industries in Baja California, Mexico. Problemas Del Desarrollo. Latin Am J Econ 51(202):63–89. https://doi.org/10.22201/iiec.20078951e.2020.202.69672
Sopelana A, Kunc M, Rivera O, Aranceta J, Rodriguez Castellanos A (2021) Modelling the dynamics of organisational change in a Spanish industrial cooperative. J Oper Res Soc 73(3):455–466. https://doi.org/10.1080/01605682.2020.1860659
United Nations (2023) Decent work and economic growth. In: Report on the sustainable development goals 2023: special edition. https://bit.ly/4b0uISk
United Nations Conference on Trade and Development (2022) Creative economy outlook 2022. https://bit.ly/404K0ib
United Nations Conference on Trade and Development (2024) Creative economy outlook 2024 in technical and statistical report. http://bit.ly/4lbphnG
United Nations Educational Scientific and Cultural Organization (2021) Cultural and creative industries in the face of COVID-19. In Overview of the economic impact. https://doi.org/10.25267/periferica.2017.i18.08
United Nations Educational Scientific and Cultural Organization (2023) The eight Creative Cities of Mexico that are part of the UNESCO International Network sign a Letter of Intent to work together in a National Network. News: UNESCO and Eight Mexican Municipalities Establish Mexican Network of Creative Cities. https://bit.ly/4bdGQPI
Vázquez-Álvarez I (2022) The book and its industry within the framework of cultural industries. In: Economic modeling and socio-economic characteristics. El Profesional de La Información, pp 1–19. https://doi.org/10.3145/epi.2022.mar.08
Wasserbaur R, Sakao T (2020) Conceptualising design fixation and design limitation and quantifying their impacts on resource use and carbon emissions. Sustainability (Switzerland) 12(19)
Wilson S (2020) Building apparel manufacturing competitiveness through policy—a system dynamics approach. J Fash Mark Manag 24(2):277–302. https://doi.org/10.1108/JFMM-10-2019-0243
Woodside AG (2016) Systems thinking and system dynamics modeling. In: Case study research, 2nd edn. Emerald Group Publishing Limited, pp 353–367. https://doi.org/10.1108/978-1-78560-461-420152029
Yang X, Qi Y (2022) Simulation of intellectual property management on evolution driving of regional economic growth. Appl Sci (Switzerland) 12(18). https://doi.org/10.3390/app12189011
Yasin NA, Ridjal S, Jufri M, Anshari (2019) Human capital and entrepreneurship and their impact on the productivity of traditional craftsmen. Espacios 40(4). https://bit.ly/4bdHgFM

Decent Work: A Transdisciplinary SDG 11

Mario Aguilar-Fernández, Graciela Salgado-Escobar, Luvis Paola León-Romero, and Joas Gómez-García

Abstract

Given the current reality, the objective of this research is to design a transdisciplinary process that promotes decent work in companies to ensure their contribution to achieving the SDGs. This is a qualitative and documentary-based study. The process is based on a literature review on the Web of Science, Wilber's holonic theory, and Checkland's soft systems methodology. It is a process comprised of four stages and four gateways, each with its central focus on companies re-envisioning, from a systemic and complex perspective, decent work as a problem for everything and everyone. Finally, it is imperative to contribute to the restoration of the social fabric and revalue work to ensure our well-being and that of the planet.

M. Aguilar-Fernández (✉) · G. Salgado-Escobar · L. P. León-Romero · J. Gómez-García
Interdisciplinary Professional Unit of Engineering and Social and Administrative Sciences of the National Polytechnic Institute, Mexico City, Mexico
e-mail: maguilarfer@ipn.mx

G. Salgado-Escobar
e-mail: gsalgadoe@ipn.mx

L. P. León-Romero
e-mail: lleonr2100@alumno.ipn.mx

J. Gómez-García
e-mail: jogomez@ipn.mx

Keywords

Interdisciplinarity · SDG 8 · Working conditions · Social dialogue

11.1 Introduction

Wars, diseases, migration, economic crises, and environmental imbalances, among other challenges currently facing the planet, are generating unprecedented, multidimensional levels of uncertainty and disruption in our lives. In more than 90% of countries, the Human Development Index has declined (UNDP 2022), and employment trends are not encouraging—a situation that does not support the reduction of the decent work deficit, which was already significant before the COVID-19 pandemic and was further exacerbated by the health crisis (ILO 2023b).

Undoubtedly, it is urgent to improve people's working conditions, since work is the sphere in which both the economic and social objectives of individuals converge. Therefore, work must guarantee protection and social security (Rodgers 2002). Unfortunately, globalized markets and political deregulation have led to transformations in the world of work: structured employment contracts have shifted to unstructured and flexible work arrangements (Eurofound 2018), with more fluid work dynamics (Touraine 2016) that offer less security and integration, as well as greater social deprivation for employees (Kalleberg and

P. Pérez-Hernandez and R. Huerta Cuervo (eds.), *Social Well-Being and the Challenges in the Sustainable Development Goals*, Sustainable Development Goals Series,
https://doi.org/10.1007/978-3-032-14294-8_10

Hewison 2013). This leads not only to precarious working conditions but also to the deterioration of individual health, family structure, and, more broadly, social life (Kalleberg and Vallas 2017).

On the other hand, the weakening and breakdown of traditional work structures observed in the contemporary labor market, in addition to making it more flexible, also render it complex and heterogeneous (Castel 2010; Touraine 2007). According to the International Labour Organization (ILO) (2016), the level of involvement of stakeholders affects, among other factors, the impact of labor regulations. Thus, actors are a variable that can influence decent work; in particular, collective and transnational actions are comparatively more successful than individual and national ones (Keohane et al. 2003; Tallberg et al. 2014). Participation leads to the successful defense of labor rights and also provides confidence and guidance to civil society actors who witness illicit labor practices (Oehri 2017).

Achieving the Sustainable Development Goals (SDGs) is therefore crucial for society, particularly Goal 8 (SDG 8), which seeks decent work for all, grounded in inclusive and sustainable economic growth, along with full and productive employment. The impact of decent work is considered to *go beyond* because it integrates economic, social, and environmental objectives (UNDP 2023). Achieving this requires social dialogue (ILO 2023a; UNDP 2023), which implies crossing borders, the participation of all stakeholders, and systemic and complex thinking.

11.1.1 Conceptual Framework

11.1.1.1 Decent Work

Throughout history, inequalities between those who hold power and those who do not have been evident. While global economies have grown, so too has social injustice (Galeano 2003); according to the ILO, in Latin America (LA) there are more than 180 million people living in extreme poverty (ILO 2023b). Mexico, like the rest of LA, has witnessed the exploitation of cheap labor, precarious working conditions, loss of employment opportunities, and wage insufficiency and inequality. This has led to countless social struggles and mobilizations to achieve recognition of labor rights. Following the Mexican Revolution of 1910, the right to decent work was recognized in Article 123 of the Mexican Political Constitution (Const 1917). However, regulations have been insufficient, as it is necessary to ensure their enforcement; thus, both the government and companies play a crucial role in achieving this.

Internationally, after the end of World War I in 1919, the ILO was established to promote social justice as a means to achieve universal and lasting peace. In the context of neoliberal hegemony, which has resulted in the loss of millions of jobs, the concept of decent work emerged in 1999 at the 87th session of the International Labour Conference as the foundation of social justice and human rights (Pouyaud 2016). Work is recognized as a source of wealth generation and social progress for countries, enabling society to move toward a dignified life (ILO 1999). Thus, decent work is defined as "productive work in which rights are protected, which generates adequate income, with appropriate social protection" (ILO 1999, p. 15). The concept encompasses four components: employment, workers' rights, social protection, and social dialogue (Ghai 2003).

Finally, in response to the ongoing historical experience of social disintegration and fragmentation (Galeano 2003), in recent decades, the decent work paradigm has served as a catalyst for the institutional action of the ILO, and has been incorporated into the United Nations 2030 Agenda for Sustainable Development. SDG 8 is proposed as an alternative to promote: the eradication of forced, child, and slave labor; employment with benefits and social security; work-life balance with equal opportunities; as well as the right to freely associate and join organizations that defend workers' interests (UNDP 2023). In companies, it is considered an indicator that demonstrates the interrelationship between corporate social responsibility and human resource development (Ardichvili 2013). Thus, the latter has proven to be a matter of fundamental values, contributing to the achievement of the common good (Kuchinke 2010).

11.1.1.2 Social Actors Involved

The social actors (Leydesdorff 2000) involved in SDG 8 are understood to include: companies, workers, higher education institutions (HEIs), government, and society, as each has a different degree of influence and interest in shaping reality. A company is understood as the integrating social institution (Drucker 1995) that emphasizes the application of regulations and supports the formation of unions in the promotion of SDG 8 (Cooke et al. 2019). Therefore, it is responsible for ensuring the progress of its stakeholders and itself (Ackoff 2000). Workers are those who possess the knowledge, skills, abilities, aptitudes, and values, and by applying them, generate value for the company (Drucker 1995). Workers have the right to unionize and engage in collective bargaining—to promote and advocate for their respective interests—through which democracy, decent working conditions, and labor market governance are fostered (ILO 2008). Higher education institutions (HEIs) are social institutions whose purpose is to train individuals capable of participating, questioning, creating, and transforming, through social conflicts and negotiations, the optimal conditions (Touraine 2016) to advance SDG 8. The government is responsible for creating and ensuring the conditions necessary to achieve the SDGs, particularly SDG 8. The community comprises all individuals who are part of society and who, individually or collectively, have the right to express themselves.

11.1.2 Applicable Theory and Methodologies

Undoubtedly, decent work requires a transdisciplinary (TD) approach, given the significance of the issue and the need to develop capacities and legitimacy. TD research is considered necessary, provided that (Scholz 2011): (i) it addresses a socially relevant problem, (ii) it facilitates mutual learning processes among actors from different disciplines and institutions, and (iii) it aims to create knowledge that is socially robust and transferable to provide solutions. Therefore, it is appropriate to draw on the TD process proposed by Wilber's holonic theory.

To ensure the transition from one stage to another, it is proposed to apply Cooper's (2008) stage-gate methodology. This methodology consists of defining stages, each comprising a series of activities that move the process to a higher level or to the starting point of the next stage. To ensure that the objectives of each stage have been met, it is necessary to establish evaluation criteria for each, that is, to define gates to move on to the next stage; otherwise, the process must be repeated.

To address the complex social issues associated with SDG 8, it is recommended to apply Checkland's (2000) Soft Systems Methodology (SSM). This methodology allows for the management of dynamic problems and the integration of different perspectives on reality. The research process is based on seven principles and consists of four actions: first, identifying and studying the problem; second, developing meaningful models to explore the problem from various perspectives or approaches; third, formulating questions around the issue, using models to identify viable and timely solutions for change; and fourth, designing and implementing actions to improve the current situation. The soft systems methodology is conceived as a continuous process in which critical reflection is fundamental.

11.1.3 Previous Research

Based on a literature review conducted in the Web of Science (WoS) on decent work, a limited number of publications were found, although there has been an increase since 2020. It is worth noting that no publications were found that directly link decent work with TD; however, one publication argues that the study of decent work should be integrated with decent labor relations and decent social relations (Cooke et al. 2019). On the other hand, some publications were identified that may be related to TD, as they consider that decent work should be promoted at various levels (Reinecke and Donaghey 2021). Others relate decent work to education (Bonvin et al. 2023; Choi et al. 2023; Di Fabio et al. 2023; Duffy et al. 2022; Kashyap et al. 2021; Kenny et al. 2023; Krishna and Singh 2005; Ouane and Singh 2005; Ribeiro et al. 2016; Rodríguez 2021).

Clearly, based on the literature review, it is necessary to design intervention strategies to ensure decent work; therefore, education and social dialogue are key to achieving this. Thus, the various authors agree that decent work is an issue that concerns governments, higher education institutions, society, and businesses, in order to guarantee equality, justice, dignity, and equity. Therefore, the objective of this research is to design a transdisciplinary process that promotes decent work in companies to ensure their contribution to achieving the SDGs. The book chapter is structured into four sections: introduction, method, results, and conclusions.

11.2 Method

This is an exploratory, qualitative, and documentary study, conducted in two stages. The first stage consists of a literature review in the Web of Science on studies that have been published on "decent work." The inclusion criteria are: (i) scientific articles and book chapters, (ii) in English, (iii) open access, and (iv) containing the term "decent work" in the title. The review is carried out in three phases (Jesson et al. 2011; Snyder 2019): (I) publications are searched for and collected using the descriptors "decent work" or "decent employment." (II) The collected publications are examined, identifying the research objective. (III) Findings are presented. Based on the results of the literature review, the second stage is developed, which is the configuration of the TD-SDG8 process for companies. This is done in three phases: (i) the stages are established, inspired by holonic theory (Wilber 1996), as well as the gates, based on the stage-gate methodology (Cooper 2008), which structure the process. (ii) Each stage of the process is described, applying SSM (Checkland 2000). And, (iii) the instrument for each gate is designed, based on the proposal on decent work by Blustein et al. (2016) and Winchenbach et al. (2019), as well as the proposal on TD competencies by Muhar et al. (2013).

11.3 Results

11.3.1 TD-SDG8 Process for Companies

The TD process to achieve SDG 8 (TD-SDG8) is a cyclical process consisting of four stages, each incorporating and transcending the previous one: disciplinary, multidisciplinary, interdisciplinary, and TD (Nicolascu 1999; Wilber 1996). Each stage is structured based on the soft systems methodology (Checkland 2000). To validate that the appropriate identity is achieved to proceed to the next stage, four gates are established (Cooper 2008). The identity (holon) is defined by the level of scope and depth with which the SDG 8 issue is studied (Wilber 1996). It is important to note that SDG 8 is composed of several dimensions and is therefore represented within a cube. It should also be noted that the TD stage not only feeds back into the disciplinary stage but is also the cornerstone. See Fig. 11.1.

11.3.1.1 Stage I. Disciplinary

Stage I consists of addressing SDG 8 in isolation, that is, each of the stakeholders works individually. Therefore, the problem is viewed from a limited level of depth and scope (Wilber 1996). From the company's perspective, based on SSM (Checkland 2000), the stage begins with the identification of the problem, for which the construction of a "rich picture" is proposed. The images should be comfortable and as natural as possible for individuals, as they will serve as the starting point for an exploratory discussion around SDG 8, in which the current situation, main stakeholders, and issues are analyzed. Subsequently, a "model" is constructed, based on declared worldviews, which should be relevant, capable of stimulating, supporting, and structuring the debate on the SDG 8 situation.

It should be noted that the premise is that companies are open systems that interact with their environment, and therefore, human activity

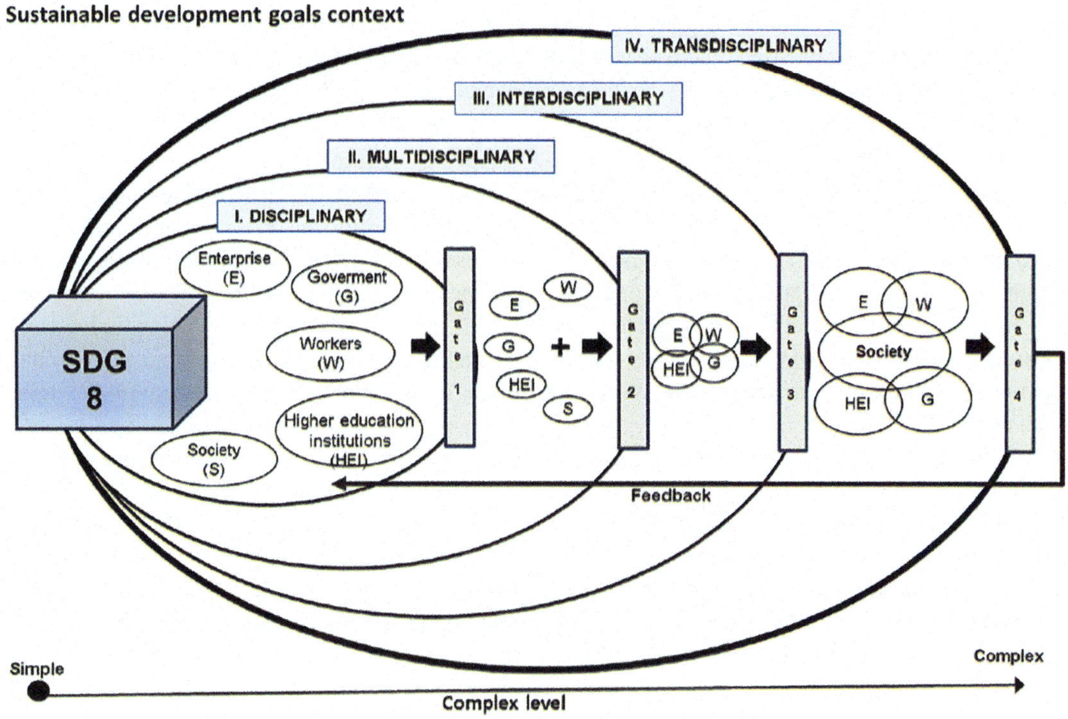

Fig. 11.1 TD-SDG8 process for companies. *Source* Own elaboration

subsystems are included as part of their modeling process (Checkland 2000). A *system* is understood as "a whole that cannot be divided into independent parts without losing its essential properties or functions" (Ackoff 2000, p. 9). Therefore, the company should "focus on the interactions of the parts and not on their actions considered separately (Ackoff 2000, p. 183)." The formation of advisory boards is relevant (see Ackoff 2000). It is expected that companies will promote new ideas and research initiatives to develop decent work and the inclusion of both vulnerable (Blustein et al. 2020) and non-vulnerable workers. This begins with job design (Bohlander et al. 2017), considering that it is an upward process carried out by employees to better adapt their own needs, aspirations, preferences, and circumstances to their jobs (Wrzesniewski and Dutton 2001). Likewise, by promoting a culture of participation and learning (Senge 2012), grounded in education for sustainable development.

Finally, at the disciplinary level, the company must comply with the fundamental rights of men and women. According to the classification of labor standards and values adopted by the ILO, which consists of three categories (see Servais 2004): (i) fundamental standards, (ii) technical standards, and (iii) programmatic standards. For this stage, the first category applies, as these are universally recognized standards, manifested in instruments such as the International Covenant on Economic, Social, and Cultural Rights, the Covenant United Nations International Covenant on Civil and Political Rights, the ILO Constitution, and the Declaration on Fundamental Principles and Rights at Work, adopted by the International Labour Conference in June 1998. These have also been incorporated into social charters by regional organizations, including NAFTA, MERCOSUR, and the European Union (Servais 2004). Thus, in order to guarantee decent work, companies must develop their codes of conduct, as well as individual and collective agreements, based on ILO declarations. In Mexico, in addition to complying with the Political Constitution of the United Mexican States and its related laws, regulations, and standards on human rights and their guarantees, companies must also adhere to labor and social welfare provisions.

11.3.1.2 Gate 1

The purpose of Gate 1 is to verify that the company is able to identify issues related to the SDG, as well as to provide sustainable solutions that promote decent work within its capacity and scope. For this purpose, an assessment tool is provided in the form of a checklist (Lewin 2011), based on the proposals of Blustein et al. (2016) and Winchenbach et al. (2019), see Table 11.1. The evaluation is carried out from two dimensions: capabilities and decent work. The first dimension considers critical awareness, proactive commitment, and participation. The second dimension assesses the individual condition of the worker and, more generally, the company. For each component of these dimensions, evidence is proposed to facilitate verification that disciplinary identity has been achieved and that the second stage of the process can begin.

11.3.1.3 Phase II. Multidisciplinarity

The multidisciplinary phase involves studying the issues related to SDG 8 from multiple disciplinary perspectives (Nicolascu 1999). That is, after each social actor has examined SDG 8 individually (from their own discipline), they must now collaborate on a single project related to the topic, applying the vision and methodology of each discipline, with the aim of generating solutions to the problem from various viewpoints (Wilber 1996). Using SSM (Checkland 2000), the first step is to assemble a working team with representatives from the different social actors; notably, the participation of the company is essential. Subsequently, the groundwork is laid to launch the project, with

Table 11.1 Checklist for the discipline of PTD-SDG8

Dimension	Component	Evidence
Capabilities	*Critical awareness*	• Open and hidden aspects of SDG 8 are identified • The causes and effects of decent work are critically analyzed
	Proactive commitment	• Willingness to work, proactive personality, and professional adaptability are demonstrated • Internal resources are leveraged individually and collectively to manage and change aspects related to SDG 8 • Individuals address and manage their life and work tasks • Professional development is promoted to facilitate workers' motivational action based on SDG 8 • Actions to promote SDG 8 are planned, implemented, and evaluated • Initiative and persistence are encouraged to help individuals achieve their goals • Empowerment of stakeholders is evident
	Participation	• Social dialogue between the company and its workers is promoted, grounded in universal rights • Relational support and contributions from company members are available • Supportive relationships are linked to progress in professional development • There is capacity to drive change
Decent work	*Worker*	• Self-respect, meaningful and satisfying work, pride, autonomy, mastery, efficiency, recognition of contribution, job satisfaction, and flourishing
	Organizational context	• Respect, equal opportunities, safe and healthy working conditions, economic security, fair remuneration, collegiality and solidarity, participation, and co-management in organizational matters
Considerations		
Depth	SDG 8 is examined from the perspective of the company and its employees	
Scope	Company	

Source Adapted from Blustein et al. (2016) and Winchenbach et al. (2019)

the goal of clarifying key terms and concepts for addressing SDG 8, as well as identifying both shared and divergent criteria. Each actor then identifies the problematic situation related to SDG 8, designs a model, and proposes solutions using their own methodologies. Finally, each actor presents solutions to the problem from their own perspective. The identity of this phase is defined by the ability of the different actors to collaborate on a single project, as well as their openness to embrace new ideas or perspectives on reality (Nicolascu 1999). In this way, the *rich picture* of the company is influenced by other perspectives, without losing its disciplinary identity.

On the other hand, according to the ILO's classification of standards, it is proposed that the second category, *technical standards*, applies to the multidisciplinary phase. This category is characterized by provisions considered technical in labor and social security legislation, in a broader sense, such as: labor administration, working and employment conditions, and social security. Currently, these issues represent the bulk of the debate due to the often conflicting interests among legislators, workers, employers, and the public interest—for example, the determination of working hours. While limited flexibility in national procedures often does not facilitate the resolution of disputes, the main obstacle is the limited capacity of the actors to recognize the diversity and divergence of viewpoints (Servais 2004).

11.3.1.4 Gate 2

To confirm that the multidisciplinary phase has been completed, the checklist shown in Table 11.2 is provided. It is important to emphasize that the identity of the second phase includes and transcends the disciplinary phase, thus establishing a multidisciplinary identity (Nicolascu 1999; Wilber 1996).

Table 11.2 Multidisciplinary checklist for PTD-SDG8

Dimension	Component	Evidence
Capabilities	*Critical awareness*	• Open and hidden aspects of SDG 8 are identified • The causes and effects of decent work are critically analyzed • The perspectives of other actors are critically analyzed
	Proactive engagement	• Willingness to work, proactive personality, and professional adaptability are demonstrated • Individuals' internal resources are leveraged both individually and collectively to manage and change aspects related to SDG 8 • People address and manage their life and work tasks • Professional development is promoted to facilitate workers' motivational action based on SDG 8 • Actions to promote SDG 8 are planned, implemented, and evaluated • Initiative and persistence are encouraged to help individuals achieve their goals • The empowerment of stakeholders is evident
	Participation	• Social dialogue among different stakeholders regarding SDG 8 is encouraged • Relational support and contributions from social actors are available • Supportive relationships are linked to progress in professional development • There is capacity to mobilize change
Decent work	*Worker*	• Self-respect, meaningful and satisfying work, pride, autonomy, mastery, efficiency, recognition of contributions, job satisfaction, and flourishing
	Organizational context	• Respect, equal opportunities, safe and healthy working conditions, economic security, fair compensation, collegiality and solidarity, participation and co-management in organizational matters
Considerations		
Depth	SDG 8 is studied from the perspectives and methodologies of the stakeholders	
Scope	Company	

Source Adapted from Blustein et al. (2016) and Winchenbach et al. (2019)

11.3.1.5 Stage III. Interdisciplinarity

The third stage involves interaction among disciplines; that is, it integrates a set of topics from various disciplines within a framework that enables the exploration of more complex issues from multiple perspectives (Holley 2017). Through interdisciplinary work, stakeholders (companies, workers, government, and HEIs) are equipped with the knowledge and skills to view the world through diverse lenses, discover the interdependence among disciplines or individual topics, and synthesize disciplines to understand broader phenomena and systems (Fortuin et al. 2013), which in this case are linked to SDG 8. Applying SSM (Checkland 2000), when the *root definitions* and the *CATWOE* of the system are developed, proposals for system improvement are formulated to address the SDG 8 issues identified in the enriched diagram. By combining and integrating methodologies, various conceptual models of the new system are constructed, compared, and evaluated in relation to the problems identified in the *rich picture* (Checkland 2000). Recommendations are also made to address the necessary changes to meet SDG 8.

For the interdisciplinary stage, the category of programmatic norms applies. These are standards that provide a regulatory framework for broader discussions about the future of work at the national, regional, and international levels (Servais 2004). Such norms are commonly established in general and flexible terms. They do not require immediate action; an example is legislation on equal opportunities and equal pay. Ultimately, these are standards designed to organize and promote the actions of all involved stakeholders.

11.3.1.6 Gate 3

To demonstrate that the interdisciplinary stage has been reached, evidence such as that shown in Table 11.3 must be provided.

11.3.1.7 Stage IV. Transdisciplinarity

The TD stage consists of *seeing beyond* (Nicolascu 1999), and therefore, the design and achievement of SDG 8 must be addressed from this perspective, in which the participation of the community, business sector, workers, government, and HEIs is essential. Thus, when applying SSM to the TD stage, it is necessary to consciously think at different levels and to tolerate that stakeholders may make different judgments about which level to adopt (Checkland 2000). Therefore, it is recommended to create a shared vision—prompted by the question "What do we want to become?"—so that the involved actors can build a common image, capable of creating a shared bond that fosters genuine commitment and gives meaning to diverse activities. However, the *shared vision* entails developing the capacity to project *visions of the future* (Senge 2012), hence the importance of each stakeholder fulfilling their role within the system, particularly HEIs. When applying SSM (Checkland 2000), issues related to SDG 8 are addressed from the current context and through the analysis of interrelations with other disciplines. Undoubtedly, systemic and complex thinking are essential for configuring the model. The TD process involves learning through social dialogue. Naturally, tensions, distrust, and disagreements arise, along with various entrenched positions that must be identified and overcome. Therefore, it is essential to be able to recognize and leverage both differences and commonalities to find solutions to problems (Senge 2012) associated with SDG 8.

On the other hand, the transdisciplinary phase is considered for developing agreements that foster economic development and growth, as well as raising living standards. Thus, TD work facilitates: cooperation among stakeholders, the enactment of legislation, and the promotion of education to ensure the achievement of SDG 8, as well as monitoring and evaluating the implementation of adopted measures to ensure policy compliance and attainment of expected outcomes. Ultimately, SDG 8 is expected to be reflected in social well-being, that is, in the *functional opportunities* and *development capabilities* that are created from the assets people possess. This is measured by the Human Development Index, whose key variables are: average income, education, and life expectancy (Sen 2012), while also ensuring equal opportunities for all members of society.

Table 11.3 Interdisciplinary checklist for the PTD-SDG8

Dimension	Component	Evidence
Capabilities	*Critical awareness*	• Both explicit and implicit aspects of SDG 8 are identified • The causes and effects of decent work are critically analyzed • The perspectives of other stakeholders are critically analyzed
	Proactive engagement	• Willingness to work, a proactive attitude, and professional adaptability are demonstrated • Individuals leverage both personal and collective internal resources to manage and effect change related to SDG 8 • People address and manage their life and work tasks • Professional development is promoted to facilitate workers' motivational action based on SDG 8 • Actions to promote SDG 8 are planned, implemented, and evaluated • Initiative and persistence are encouraged to help individuals achieve their goals • Stakeholder empowerment is made visible • Systemic and complex thinking is developed • Knowledge is strengthened through the interrelation of stakeholders • It is assumed that solving issues related to SDG 8 depends on exchange and cooperation with other disciplines
	Participation	• Social dialogue among different stakeholders regarding SDG 8 is promoted • Relational support and contributions from social actors are available • Supportive relationships are linked to progress in professional development • There is capacity to mobilize change • Responsibility and commitment in shaping the social fabric are recognized
Decent work	*Worker*	• Self-respect, meaningful and satisfying work, pride, autonomy, mastery, efficiency, recognition of contribution, job satisfaction, and flourishing
	Organizational context	• Respect, equal opportunities, safe and healthy working conditions, economic security, fair compensation, collegiality and solidarity, participation and co-management in organizational matters
Considerations		
Depth	SDG 8 is studied through the integration of stakeholders' perspectives and methodologies	
Scope	Company	

Source Adapted from Blustein et al. (2016) and Winchenbach et al. (2019)

11.3.1.8 Gate 4

The final gate, while not more important than the previous ones, is nonetheless essential. This is because the TD-SDG8 process is cyclical; therefore, the TD identity is what initiates a new process: disciplinary, multidisciplinary, interdisciplinary, and TD, and the cycle begins again. Thus, it is crucial to validate that the TD phase has been achieved. See the checklist in Table 11.4.

11.4 Conclusions

The TD-SDG8 process focuses on the integration and participation of social actors, among whom companies and their workers are fundamental. The TD-SDG8 process is structured around the four stages of TD: discipline, multidiscipline, interdiscipline, and TD. At each stage, a new identity is formed, which enables transition to the next level. To verify that the identity has been acquired, the process includes a gateway between each stage. Although there are studies on decent work, publications on the subject remain limited, despite its status as an SDG. Few studies propose approaches to address decent work through the participation of social actors, and none directly link decent work with TD and the application of SSM.

Although the TD-SDG8 process is a theoretical proposal—based on a literature review in WoS—it is presented more as a culture than a process, aimed at addressing the problems

Table 11.4 Transdisciplinary checklist for the TD-SDG8 process

Dimension	Component	Evidence
Capabilities	*Critical awareness*	• Open and hidden aspects of SDG 8 are identified • The causes and effects of decent work are critically analyzed • The perspectives of other stakeholders are critically analyzed • There is an ability to see the *whole* and its constituent parts • Reality is understood and explained • There is an understanding of who we are and what constitutes us as human beings
	Proactive commitment	• Willingness to work, a proactive personality, and professional adaptability are evident • Individuals' internal resources are leveraged both individually and collectively to manage and change aspects related to SDG 8 • People address and manage their life and work tasks • Professional development is promoted to facilitate workers' motivational action based on SDG 8 • Actions to promote SDG 8 are planned, implemented, and evaluated • Initiative and persistence are encouraged to help people achieve their goals • The empowerment of stakeholders is evident • Systemic and complex thinking is developed • Knowledge is strengthened through the interrelation of stakeholders • It is assumed that solving problems related to SDG 8 depends on exchange and cooperation with other disciplines, as well as the current context
	Participation	• Social dialogue among all stakeholders involved in SDG 8 is promoted • Relational support and the contribution of social actors are available • There is the capacity to drive change • Responsibility and commitment are understood as shaping the social fabric • Knowledge is conceived as being generated collectively
Decent work	*Worker*	• Self-respect, meaningful and satisfying work, pride, autonomy, mastery, efficiency, recognition of contribution, job satisfaction, and flourishing
	Organizational context	• Respect, equal opportunities, safe and healthy working conditions, economic security, fair compensation, collegiality and solidarity, participation and co-management in organizational matters
	Socioeconomic and political context	• The prestige and social status of employment, equality, economic growth as a means to achieve decent work, minimum wage legislation, moral economy, and decent work recognized and promoted through international agendas and policies
Considerations		
Depth	SDG 8 is studied from a social context and through the integration of perspectives and methodologies from all involved stakeholders	
Scope	Society	

Source Adapted from Blustein et al. (2016), Winchenbach et al. (2019), and Muhar et al. (2013)

arising from indecent work; a culture that must be established within companies. However, the research is limited to being only an analytical proposal from a business perspective. Future research should further explore the perspectives and action processes of each stakeholder involved, as well as implement the process for validation and improvement. Unfortunately, there are still deeply ingrained traits in society, particularly within companies, that seem to have universal validity and have become so common that it is difficult to even imagine the possibility of them being otherwise (Douglas 1969), such as job insecurity and labor injustice. Therefore, it is imperative to promote a culture of social well-being that is not subordinated to particular interests (Sen and Kliskerberg 2007). It is everyone's responsibility to ensure decent work, and thus, legislation and social policies must be redesigned to address current and future economic needs, but even more so, social needs (Servais 2004).

Finally, the TD-SDG8 process aims for companies to reconceive, from a systemic and complex perspective, decent work as an issue that concerns everyone and everything. Thus, the intention of this work is to contribute to the promotion of integration and participation in society, and above all, to universal rights—justice, equality, equity, and dignity—to achieve the goals of sustainable development, particularly SDG 8.

References

Ackoff R (2000) Recreación de las corporaciones: Un diseño organizacional para el siglo XXI. Oxford University Press

Ardichvili A (2013) The role of HRD in CSR, sustainability, and ethics: a relational model. Hum Resour Dev Rev 12:456–473. https://doi.org/10.1177/1534484313478421

Blustein D, Olle C, Connors-Kellgren A, Diamonti A (2016) Decent work: a psychological perspective. Front Psychol 7. https://doi.org/10.3389/fpsyg.2016.00407

Blustein D, Perera H, Diamonti A, Gutowski E, Meerkins T, Davila A, Erby W, Konowitz L (2020) The uncertain state of work in the US: profiles of decent work and precarious work. J Vocat Behav 122. https://doi.org/10.1016/j.jvb.2020.103481

Bohlander G, Snell S, Morris S (2017) Administración de recursos humanos. Cengage Learning

Bonvin J, Cianferoni N, Mexi M (2023) Conclusion: the rise and growth of the gig economy. In: Bonvin J, Cianferoni N, Mexi M (eds) Challenges and opportunities for social dialogue and decent work (WOS:001010827800007), p 158. https://doi.org/10.4337/9781800372375

Castel R (2010) EL ASCENSO DE LAS INCERTIDUMBRES. Trabajo, protecciones, estatuto del individuo. Fondo de Cultura Económica

Checkland P (2000) Soft systems methodology: a thirty year retrospective. Syst Res Behav Sci 17:48

Choi S, Li H, Ogawa K (2023) Upper secondary vocational education and decent work in Indonesia: a gender comparison. Int J Educ Dev 101. https://doi.org/10.1016/j.ijedudev.2023.102833

Constitución Política de los Estados Unidos Mexicanos [Const] Art. 123. 5 de febrero de 1917 (México)

Cooke F, Xu J, Bian H (2019) The prospect of decent work, decent industrial relations and decent social relations in China: towards a multi-level and multi-disciplinary approach. Int J Hum Resour Manag 30(1):122–155. https://doi.org/10.1080/09585192.2018.1521461

Cooper R (2008) Perspective: the stage-gates idea-to-launch process—update, what's new, and NexGen systems. J Prod Innov Manag 25:213–232. https://doi.org/10.1111/j.1540-5885.2008.00296.x

Di Fabio A, Svicher A, Kenny M (2023) A network analysis of relationships between the work as meaning inventory (WAMI-U) and the humor styles in university students: enhancing prevention in decent education. Aust J Career Dev 32(3):203–214. https://doi.org/10.1177/10384162231208641

Douglas M (1969) El aspecto humano de las empresas. Diana

Drucker PF (1995) Managing in a time of great change

Duffy RD, Kim HJ, Perez G, Prieto CG, Torgal C, Kenny ME (2022) Decent education as a precursor to decent work: an overview and construct conceptualization. J Vocat Behav 138:103771. https://doi.org/10.1016/j.jvb.2022.103771

Eurofound (2018) Non-standard forms of employment: recent trends and future prospects. Publications Office of the European Union

Fortuin K, Van Koppen C, Kroeze C (2013) The contribution of systems analysis to training students in cognitive interdisciplinary skills in environmental science education. J Environ Stud Sci 3:139–152. https://doi.org/10.1007/s13412-013-0106-3

Galeano E (2003) Las venas abiertas de América Latina. Siglo XXI

Ghai D (2003) Decent work: concept and indicators. Int Labour Rev 142(2):113. https://doi.org/10.1111/j.1564-913X.2003.tb00256.x

Holley KA (2017) Interdisciplinary curriculum and learning in higher education. https://api.semanticscholar.org/CorpusID:157462952

Jesson J, Matheson L, Lacey FM (2011) Doing your literature review: traditional and systematic techniques. SAGE

Kalleberg LA, Vallas PS (eds) (2017) Probing precarious work: theory, research, and politics. In: Precarious work, vol 31. Emerald Publishing Limited, pp 1–30. https://doi.org/10.1108/S0277-283320170000031017

Kalleberg AL, Hewison K (2013) Precarious work and the challenge for Asia. Am Behav Sci 57(3):271–288. https://doi.org/10.1177/0002764212466238

Kashyap V, Nakra N, Arora R (2021) Do "decent work" dimensions lead to work engagement? Empirical evidence from higher education institutions in India. Eur J Train Dev 46(1/2):158–177. https://doi.org/10.1108/EJTD-08-2020-0127

Kenny M, Schmidtberger R, Masters A (2023) Promoting decent work and decent life for all: preparing the next generation through decent education and career development education. Aust J Career Dev 32(3):187–195. https://doi.org/10.1177/10384162231186115

Keohane R, Moravcsik A, Slaughter A-M (2003) Legalized dispute resolution: interstate and transnational. Int Org 54. https://doi.org/10.1162/002081800551299

Krishna V, Singh M (2005) Education, training and skills formation for decent work in the informal sector: case studies from northern India. In: Meeting basic learning needs in the informal sector: integrating education and training for decent work, empowerment and citizenship, vol 2, pp 183–214. https://doi.org/10.1007/1-4020-3427-X

Kuchinke K (2010) Human development as a central goal for human resource development. Hum Resour Dev Int 13:575–585. https://doi.org/10.1080/13678868.2010.520482

Lewin E (2011) El efecto Checklist. Como Una Simple Lista De Comprobacion Elimina Errores Y Salva Vidas. Antoni Bosch Editor

Leydesdorff L (2000) The triple helix: an evolutionary model of innovations. Res Policy 29(2):243–255. https://doi.org/10.1016/S0048-7333(99)00063-3

Muhar A, Visser J, van Breda J (2013) Experiences from establishing structured inter- and transdisciplinary doctoral programs in sustainability: a comparison of two cases in South Africa and Austria. J Clean Prod 61:122–129. https://doi.org/10.1016/j.jclepro.2013.07.031

Nicolascu B (1999) La evolución transdisciplinaria del aprendizaje. Trans-Pasando Fronteras 4:39–50. https://doi.org/10.18046/retf.i4.1779

Oehri M (2017) Civil society activism under US free trade agreements: the effects of actorness on decent work. Polit Govern 5(4):40–48. https://doi.org/10.17645/pag.v5i4.1085

ILO (1999) Trabajo decente. Conferencia Internacional del Trabajo 87.a reunión 1999

ILO (2008) La libertad de asociación y la libertad sindical práctica: Lecciones extraidas. https://www.ilo.org/global/topics/freedom-of-association-and-the-right-to-collective-bargaining/lang--es/index.htm

ILO (2016) Assessment of labour provisions in trade and investment arrangements [Report]. http://www.ilo.org/global/publications/books/WCMS_498944/lang--en/index.htm

ILO (2023a) La ecologización de las empresas: Transformar los procesos y los lugares de trabajo (1a ed.). Organizacion Internacional del Trabajo. https://doi.org/10.54394/QNGU7279

ILO (2023b) Perspectivas Sociales y del Empleo en el Mundo: Tendencias 2023. Organizacion Internacional del Trabajo. http://www.ilo.org/global/research/global-reports/weso/WCMS_881012/lang--es/index.htm

Ouane A, Singh M (2005) Meeting basic learning needs in the informal sector integrating education and training for decent work, empowerment and citizenship preface. In: Meeting basic learning needs in the informal sector: integrating education and training for decent work, empowerment and citizenship (WOS:000270210900002), vol 2, pp X–XI. https://doi.org/10.1007/1-4020-3427-X

Pouyaud J (2016) For a psychosocial approach to decent work. Front Psychol 7. https://doi.org/10.3389/fpsyg.2016.00422

Reinecke J, Donaghey J (2021) Towards worker-driven supply chain governance: developing decent work through democratic worker participation. J Supply Chain Manag 57(2):14–28. https://doi.org/10.1111/jscm.12250

Ribeiro MA, Silva FF, Figueiredo PM (2016) Discussing the notion of decent work: senses of working for a group of Brazilian workers without college education. Front Psychol 7. https://www.frontiersin.org/articles/10.3389/fpsyg.2016.00207

Rodgers G (2002) El trabajo decente como una meta para la economía global. CINTERFOR/OIT

Rodríguez J (2021) The promotion of decent work through the cooperative principle concerning education, training and information. Boletin de la Asociacion Internacional de Derecho Cooperativo-Int Assoc Cooper Law J 58:115–135. https://doi.org/10.18543/baidc-58-2021pp115-135

Scholz R (2011) Environmental literacy in science and society. From knowledge to decisions. Raumforschung und Raumordnung 71(1):69–71. https://doi.org/10.1007/s13147-012-0203-x

Sen A, Kliskerberg B (2007) Primero la gente: Una mirada desde la ética del desarrollo a los principales problemas del mundo globalizado. Grupo Planeta

Sen A (2012) La idea de la justicia. TAURUS

Senge PM (2012) La quinta disciplina (Nueva Edición): El arte y la práctica de la organización abierta al aprendizaje

Servais J-M (2004) Globalization and decent work policy: reflections upon a new legal approach. Int Labour Rev 143(1–2):185–207. https://doi.org/10.1111/j.1564-913X.2004.tb00550.x

Snyder H (2019) Literature review as a research methodology: an overview and guidelines. J Bus Res 104:333–339. https://doi.org/10.1016/j.jbusres.2019.07.039

Tallberg J, Sommerer T, Squatrito T, Jönsson C (2014) Explaining the transnational design of international organizations. Int Org 68(4):741–774. Cambridge Core. https://doi.org/10.1017/S0020818314000149

Touraine A (2007) New paradigm for understanding today's world. Polity Press

Touraine A (2016) El fin de las sociedades. Fondo de Cultura Economica

UNDP (2022) Human development report 2021–22. Programa de las Naciones Unidas para el Desarrollo. http://report.hdr.undp.org

UNDP (2023) Objetivo 8: Trabajo decente y crecimiento económico | Objetivos de Desarrollo Sostenible | Programa De Las Naciones Unidas Para El Desarrollo. UNDP. https://www.undp.org/es/sustainable-development-goals/trabajo-decente-crecimiento-economico

Wilber K (1996) Breve historia de todas las cosas. Editorial Kairós

Winchenbach A, Hanna P, Miller G (2019) Rethinking decent work: the value of dignity in tourism employment. J Sustain Tour 27(7):1026–1043. https://doi.org/10.1080/09669582.2019.1566346

Wrzesniewski A, Dutton JE (2001) Crafting a job: revisioning employees as active crafters of their work. Acad Manag Rev 26(2):179–201

Sustainability of Communal Enterprises in the New Institutional Economics in Mexico

12

Juan Regino Maldonado, Marco Ricardo Téllez Cabrera, Verónica Hernández Mendoza, and María Guadalupe Obregón Sánchez

Abstract

The research examines forest sustainability in the timber production of communal enterprises of the new institutional economics (NIE), from the perspective of socio-ecological systems (SESs), considering 8 categories or first-order variables, of which 6 are internal and 2 are external to the forest system. The research was designed under the methodological requirements of case studies on two enterprises. The Communal Forestry and Agricultural Services Unit of Ixtlán de Juárez, Oaxaca (UCFAS by its acronym in Spanish) and the Forestry Utilization and Integral Development Unit of Nuevo San Juan Parangaricutiro, Michoacán (UAFyDI by its acronym in Spanish). Through a structured questionnaire applied to the owners of the enterprises in 2023, the information from the categories included in the research is collected, recorded, and qualitatively analyzed. The study finds that forest sustainability in both cases is related to the interaction between the ecological and social systems present in the communities. It is concluded that the communal forest enterprises have advanced toward forest sustainability mainly related to forest governance with positive implications on the Agenda of the Sustainable Development Goals (SDGs).

Keywords

New institutional economics · Socio-ecological systems · Communal enterprises · Community forest enterprises · Sustainability

12.1 Introduction

In the last decades of the twentieth century, the institutional approach regained importance because, through the new institutional economics(NIE), a theoretical framework was constructed that places institutions in the mainstream economic discourse and incorporates concepts from social, environmental, business organization, history, and law sciences. The NIE

J. R. Maldonado (✉)
Instituto Politécnico Nacional CIIDIR Unidad Oaxaca, Santa Cruz Xoxocotlán, Oaxaca, Mexico
e-mail: jregino@ipn.mx

M. R. T. Cabrera
Sección de Estudios de Posgrado e Investigación-Escuela Superior de Economía, Instituto Politécnico Nacional, Mexico City, Mexico
e-mail: mtellezc@ipn.mx

V. H. Mendoza
Universidad Nacional Rosario Castellanos, Gustavo A. Madero, Mexico City, Mexico
e-mail: veronica.hernandez@rcastellanos.cdmx.gob.mx

M. G. O. Sánchez
Unidad Profesional Interdisciplinaria de Ingeniería y Ciencias Sociales y Administrativas del Instituto Politécnico Nacional, Iztacalco, Mexico City, Mexico
e-mail: mobregon@ipn.mx

P. Pérez-Hernandez and R. Huerta Cuervo (eds.), *Social Well-Being and the Challenges in the Sustainable Development Goals*, Sustainable Development Goals Series,
https://doi.org/10.1007/978-3-032-14294-8_11

aims to clarify the role that institutions play in the design and implementation of the rules of the game for the management of productive, technical, mainly natural resources, and how these impact the economy, population, and environment (Alston 2018; Vega and Keenan 2016).

In the economics, formal and informal institutions establish rules, norms, and incentives for the creation and proliferation of conventional enterprises. However, in rural areas, there is empirical evidence that the institutions that make up the community governance system based on autonomy, self-management, and self-organization also establish rules (operational, collective and constitutional), norms, incentives, and sanctions to create and develop communal enterprises (Loreggian et al. 2023; Merino-Pérez 2018; Ostrom 2000; Garibay 2007).

Communal enterprises, mainly in rural areas, emerge in an open economic context with economic crisis, inflation, environmental degradation, and poverty (Peredo and Chrisman 2006). Typically emerging in rural areas, in peasant and indigenous communities, their activities are related to the production, transformation, and marketing of sustainable agricultural and forestry products; in this sense, they respond with proposals on alternative production systems to positive economics, surviving in market niches with low market shares, but with certain income (Oberlack et al. 2015; Engbring and Hajjar 2021; Cubbage 2015).

Communal enterprises of the new institutional economics(CENIE)—the theoretical perspective adopted in this research, given the institutional arrangement presented by the social and productive organization system of these enterprises, can be studied from this approach—emerge as a response to various problems faced by communities settled in rural areas. They have the particularity of organizing around communal land ownership, their own forms of organization, and common objectives (Toledo and Ortíz-Espejel 2014). For example, the slow and limited development of economic activities and the persistence of social deficiencies in the population can lead to an increase in emigration, diseases, and consequently, the mortality of vulnerable populations (Abrahams and Peredo 1996). Exacerbating their problems, uncontrolled and unregulated anthropogenic activities such as changes in land use due to intensive monoculture practices and population growth, expansion of the agricultural and livestock frontier (SEMARNAT 2003; Gómez-Mendoza et al. 2006). Due to climate change, some communities face serious environmental problems such as loss of forest cover, soil loss, reduction and contamination of water bodies, among the most mentioned in the literature (Azqueta et al. 2007).

In Mexico, as in other countries, one way to address environmental problems was to return the lands to the peasants (White and Martin 2002; Zhang et al. 2000). In 1992, with the reform of Article 27 of the Constitution, land ownership was regulated as private, ejidal, and communal; the latter two known as social property (Muñoz-Piña et al. 2003; Madrid et al. 2009). While ejidal property allows the land to be subdivided into private lots and sold, communal property only allows the land to be subdivided for use, but not sold (Agrarian Law 2002). Changes in agrarian legislation encouraged the creation of forestry enterprises initially known as "work groups" made up of individual owners, under the legal category of Rural Farmers' Production Societies (Wilshusen 2003). Currently called community forestry enterprises (CFEs) (Antinori and Bray 2005), which in this study we refer to as CENIE. In this sense, these enterprises have tried to contribute to the resolution of environmental problems in communities by developing community governance systems to strengthen collective action in biodiversity conservation and sustainable use of natural resources for local development based on natural sustainability (Tetreault 2004). Therefore, the positive contribution of this business model to the Agenda of the SDGs can be proposed. For example, in addition to the production and marketing of wood, some of these enterprises generate income from the sustainable use of non-timber forest products, ecotourism, water capture and carbon, conservation of wild flora and fauna, among others (Engbring and Hajjar 2021; 2022; Zhang et al. 2021; Toledo et al. 2014). Due to government support through programs such as the program environmental services (PES) and recently "Sembrando Vida" in Spanish (Fuente and Marin 2024; Cubbage et al. 2015). Therefore, in this research with a case study design, the forest sustainability of timber production of two communal enterprises located in Oaxaca and Michoacán is examined from the perspective of the NIE based on the SSEs framework (McGinnis and Ostrom 2014).

12.2 CENIE

This business model arises from the collective initiative, productive, and environmental experience of communities for sustainable local and regional economic development (Bray et al. 2006; Antinori and Bray 2005). With broad possibilities to positively contribute to achieving several SDGs, particularly 15, on life of terrestrial ecosystems where sustainable forest management under the responsibility of communities is paramount in sustainable development (Tetreault 2004). However, to analyze the CENIE, it is necessary to address the emergence of this type of enterprise, identify the necessary conditions and at the same time, the key factors enhancing their growth. Finally, identify the distinctive features they present compared to conventional enterprises.

12.2.1 Emergence of the CENIE

In Mexico between 1940 and 1970, the government authorized concessions to private and parastatal enterprises for forest extraction, which contributed to the proliferation of sawmills and illegal logging (Bray et al. 2006). Between 1974 and 1986, several of the forest extraction concessions granted by the state to private and parastatal enterprises expired, a situation that was taken advantage of by the communities to begin the fight for forest rights. In this way, in Oaxaca, a group of farmers emerged known as the Organization in Defense of Natural Resources and Social Development of the Sierra Juárez (ODRENASIJ by its acronym in Spanish), who proposed the restoration of forest rights in favor of CFEs, and not in the hands of private enterprises and the state (Merino-Pérez and Segura-Warnholtz 2005). In 1986, the Law ended the concessions to private enterprises for forest extraction and began the dismantling of parastatal forest enterprises. Since then, communities have obtained land ownership rights and, therefore, the management of forest resources for the extraction of timber (Bray et al. 2006). The 1997 Law established the support program for the development of commercial forest plantations (PRODEPLAN by its acronym in Spanish) and the forest development program (PRODEFOR by its acronym in Spanish). The first program aimed to develop productive projects, and the second, to promote the sustainable extraction of natural ecosystems. In 2001, the program for the conservation and sustainable management of forest resources (PROCYMAF by its acronym in Spanish) was created, with the aim of promoting the strengthening of ejido and community organizations, as well as the creation and operation of CFEs, which in this study are referred to as CENIE. In 2009, the same program created the framework to classify them into 4 types, according to the development of their organizational capabilities for the management of natural resources, capitalization, and vertical integration in the forest production chain (Segura-Warnholtz 2014). Type I enterprises, those that have forest resources but do not carry out management activities; Type II enterprises, those that sell their logging rights to private or communal enterprises; Type III enterprises have some forest lands and sell their roundwood, and Type IV enterprises are those that transform forest resources and add value to timber products. From 2011 to the present, this same program was renamed the community forest development program, which promoted the technical training of forest advisors, community organization processes, and continues to support the CFEs (CONAFOR 2013).

12.2.2 Necessary Conditions and Growth Enhancers of the CENIE

Communities understood in the space—collectivity relationship must have at least four necessary conditions for the emergence and flourishing of the CENIE: (i) common-pool resources, (ii) governance system, (iii) social capital, and (iv) development of technological production capabilities.

(i) **Common-Pool Resources**

Large common-pool resources such as oceans and atmosphere are difficult to manage due to the difficulty of establishing limits and property rights (Ostrom and Hess 2007). In this sense, communities with small- and medium-sized pool resources have an advantage over the larger ones. Empirical evidence shows that the sustainability of natural common pool resources in communal property is possible when property rights in communities are clear (access, extraction, management, exclusion, and alienation), and institutional design principles (clearly defined boundaries, coherence between appropriation and provision rules with local conditions, collective choice arrangements, supervision, graduated sanctions, conflict resolution mechanisms, minimum recognition of organizational rights, and nested entities) (Jurjonas et al. 2023; Ostrom 2000; Bassi and Carestiato 2016).

(ii) **Governance System**

The governance structure in communities for the extraction of common-pool resources is an integration of the indigenous normative system (system of uses and customs represented by the General Assembly of Commoners (GAC)), with the business system of transformation of natural resources (Bray et al. 2006). Under vertical hierarchy, the highest authority is the GAC, which has territorial control and autocratic control of resources and the enterprise (Garibay 2007).

(iii) **Social Capital**

The organization and active involvement of community members is essential for achieving short-, medium-, and long-term objectives and goals set by the community through coordinated and collective actions (Ostrom 1998). This participation ranges from collective work to conserve or extract natural areas to collective organization for the defense of territory and natural resources.

(iv) **Development of Technological Production Capabilities**

Evidence shows that the productive activities developed by the community are related to the skills and collective experiences acquired by the members of the community before the creation of the enterprise (Peredo and Chrisman 2006; Peredo and McLean 2006). The population of indigenous communities possesses and passes on to future generations ancestral knowledge, which forms part of the culture and worldview of native peoples, related to cultivating the land and sustainable extraction of natural resources. In Mexico since the forties, the farmers as employees of parastatal and private enterprises have developed skills and abilities in the extraction and transformation of timber. The 1986 Law, which transferred the property rights of land and forest resources to the farmers, and the reform to article 27 of the constitution in 1992 forced the commoners to develop technological capabilities to produce timber. Since 2001, programs have been developed to strengthen the management, institutional, and technological capabilities of the enterprises (Cubbage et al. 2015).

12.2.3 Distinctive Features of the CENIE

The CENIE based on communal property and management of common-pool resources (Agrawal 2001; Engbring and Hajjar 2021, 2022; Regino et al. 2017). The owners with a humanistic business ethic (Zhang et al. 2021), have the right to extract natural resources, they are families that make up a corporate economic unit based on values of autonomy, self-organization, trust, cooperation, and reciprocity (Ostrom 2009; Garibay 2007). In order of importance, they generally have the following three objectives: sustainability of common-pool resources, social welfare, and economic benefits (Lueck 1994; Regino et al. 2017).

According to Garibay (2007), in Mexico the CFEs or the CENIE are the result of five features that characterize them: (i) the members are descendants of farmers from indigenous cultures of small communities; (ii) they own large extensions of communal territory and forest resources that are managed and extracted by a single economic unit for forest extraction; (iii) they have a centralized and hierarchical administration; they organize the extraction and transformation of wood into timber products that are sold in local, regional, and global market networks; (iv) they develop strong local systems of self-government, and self-organization with a certain

Table 12.1 Distinctive sustainability features of CENIE

Features	CENIE	Positive economic enterprises	Social and solidarity economic enterprises
Owner	Community members who are entrepreneurs and workers at the same time	Entrepreneur, shareholders, private cooperative partners	Entrepreneur, private cooperative partners
Main objective	Sustainability of common pool resources	Maximization of profits	Social welfare
Secondary objectives	Social welfare Economic welfare	Competitiveness	Environmental care Economic welfare
Distribution of benefits	Profits for community members Reinvestment in physical and productive capital Social infrastructure Culture Sustainability of common pool resources	Entrepreneur, shareholder, partners	Entrepreneur, partners, community, and environment
Values	Autonomy Self-management Trust Cooperation Reciprocity	Leadership Individualism	Autonomy Self-management Solidarity Cooperation
Land tenure	Communal	Private	Private Ejidal
Property rights	Access Extraction Management Exclusion Alienation (No sale-transfer)	Access Extraction Management Exclusion Alienation (Sale)	Access Extraction Management Exclusion Alienation (Sale)
Organizational structure	Hierarchical-vertical-autocratic Indigenous normative systems Business system	Hierarchical-horizontal-democratic	Hierarchical-vertical-autocratic Indigenous normative systems Business system
Productive activities	Timber forest products Non-timber forest products Ecotourism Biodiversity Conservation Water capture Carbon capture	Agroindustry	Agriculture Livestock Artisanal Cultural Food and beverage production

Source Prepared by the authors based Ostrom (2000), Garibay (2007), Peredo and Chrisman (2006), Barkin et al. (2009), Hernández et al. (2020)

degree of autonomy from external powers; and (v) they construct the community ethos of values and practices according to their social order. Table 12.1 presents the main distinctive features of CENIE that distinguish it from conventional enterprises and the model of social and solidarity economicenterprises.

12.3 Sustainability of the CENIE

Recently, the NIE has delved into the institutional approach for the management of common pool resources (Caballero and Garza 2010; Leslie et al. 2015). An example of this is the academic work of Elinor Ostrom regarding the role of governance of common pool resources, and the importance of developing a framework for assessing the sustainability of socio-ecological systems (SESs) (McGinnis and Ostrom 2014; Ostrom 2009, 2000). Where the social system is interdependent with human organisms, while the ecological system is interdependent with biophysical organisms or units (Berkes and Folke 1998). That is, both contain organisms that interact interdependently, and may contain individually interactive subsystems. Anderies et al. (2004) define SESs as a subset of social systems in which some of the interdependent relationships between human organisms are mediated through interactions with non-human biophysical units. Ostrom developed a framework of social and ecological variables for the analysis of the sustainability of SESs (McGinnis and Ostrom 2014). In Fig. 12.1, Ostrom proposes to focus the analysis on 8 categories or first-order variables, of which 6 are internal: Focal action situations composed of interactions (I) and outcomes (O); resource system (RS), resource units (RU), governance system (GS), and actors (A)), and 2 are external to the system (social, economic, and political environment (SEPE) and related ecosystems (RE)). In the focal action situations, the multi-linked interactions between the 4 main variables of the framework (RS, RU, GS, and A) take place; on management practices and different aspects related to the preservation, conservation, extraction, and sustainability assessment of common pool resources. In which, the impact and modifica-

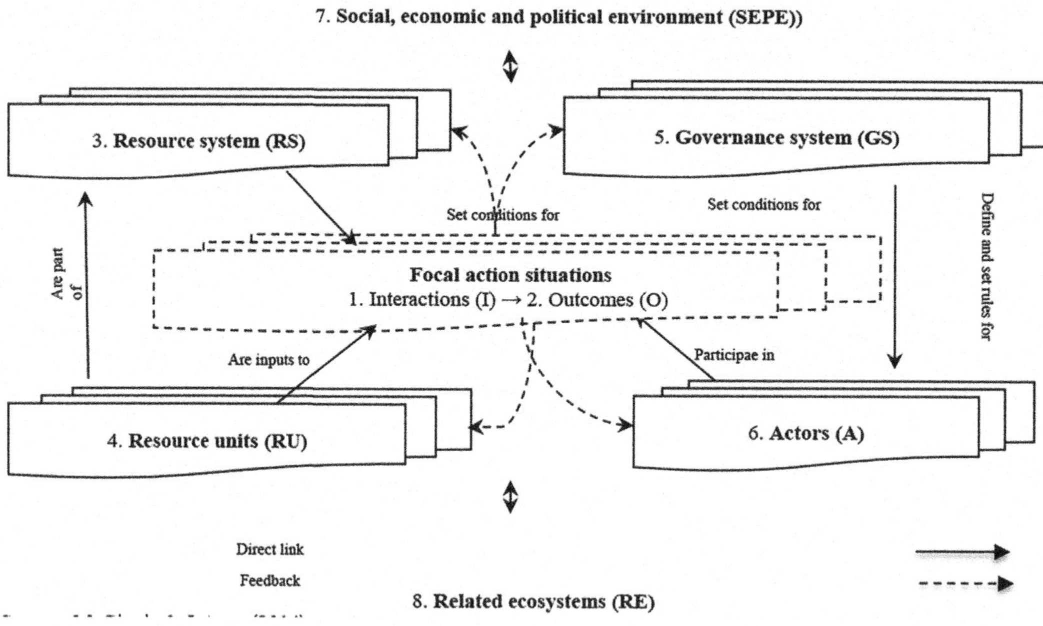

Fig. 12.1 Sustainability of SSEs. *Source* McGinnis and Ostrom (2014)

tion of the outcomes from external factors (SEPE, RE) are also considered.

12.4 Methodology

12.4.1 Type of Study

As this is an exploratory but in-depth study of the sustainability of the CENIE in Mexico. It begins with a qualitative study designed under the requirements of a case study (Caulley and Dowdy 1987; Yin n.d.). As indicated in Table 12.2, based on the SESs framework (McGinnis and Ostrom 2014), out of a total of 8 categories, 22 dimensions with 48 indicators are identified, of which only information on some of these was obtained, which is presented in the results. However, they proved quite useful for analyzing the forest sustainability of timber production of the two case studies in 2023.

12.4.2 Measurement Instrument, Information Analysis, and Results Presentation

A structured questionnaire in two parts was designed as an instrument for data collection, targeting community members and enterprise directors. The first part collected socioeconomic information about the enterprise (enterprise name and address, municipality, number of community members, education, gender and age of community members, main products, and timber volumes). The second part contains specific questions about the 48 sustainability indicators. The study is complemented with qualitative information from community visits and direct observation (Flick 2007), open interviews with key informants from the enterprises (Guber 2011). The information obtained from the main indicators was concentrated in each of the 22 dimensions that correspond to each of the 8 categories of sustainability analysis of the enterprises. The qualitative research evaluated its scientific rigor through different criteria, such as saturation that were validated in the methodological procedure of the study (Palacios-Izcara 2014; Denzin and Lincoln 2012). Finally, the results are presented by enterprise cases in two sections. The first presents socioeconomic information about the enterprise, and the second presents the main results of each of the 8 categories.

12.5 Results

12.5.1 Emergence of the UCFAS

The UCFAS is located in the municipality of Ixtlán de Juárez, belonging to the Central-North Region of the state of Oaxaca, in the foothills of the Sierra Madre known as Sierra Juárez. The communities of Santiago Textitlán, Ixtlán de Juárez, and Pueblos Mancomunados located in the Sierra Sur and Norte de Oaxaca are grouped in the Integradora Comunal Forestal de Oaxaca, S.A. de C.V. (ICOFOSA by its acronym in Spanish); with a trajectory of more than 25 years, they are evidence that organized communities achieve sustainable schemes for the management of common extraction forest resources. The integrator promotes the manufacture, industrialization, and marketing of wooden furniture. The Communal Forestry Agricultural and Services Unit (UCFAS by its acronym in Spanish) of Ixtlán de Juárez is one of the economic units of ICOFOSA, which started its operations in July 1988, with the objective of generating added value to wood through sustainable industrial transformation (Regino et al. 2015).

12.5.1.1 Forest Sustainability

Focal action situations:

1. **I.** *1. Production, transformation, and sale of wood*. During the 24 months of the study, they extracted 15,328.86 m^3 of roundwood, from which they obtained 7,917.05 m^3 of sawn wood. A portion of the wood was sold to the construction sector, and the other part was used in the manufacture of furniture for homes, schools, and offices. This wood went

Table 12.2 Forest sustainability of UCFAS-Oaxaca and UAFyDI-Michoacán

Category	Dimension	Indicator
1. Interactions (I)	1. Production, transformation, and sale of timber	1. Roundwood (m^3)
		2. Sawn wood (m^3)
		3. Kiln-dried wood (m^3)
		4. Finger joint board (m^2)
2. Outcomes (O)	2. Forest sustainability	5. Certified wood (m^3)
3. Forest Resource System (FRS)	3. Forest resource extraction limits	6. Area for timber extraction (ha)
	4. Size of the forest resource system	7. Total forest area (ha)
4. Forest Resource Units (FRU)	5. Tree species growth	8. Tree height (m)
		9. Tree crown diameter (m)
		10. Harvested pine volume (m^3)
	6. Economic values	11. Technological assets ($)
		12. Finished products ($)
5. Forest Governance System (FGS)	7. Governmental organizations	13. Number of federal organizations
		14. Number of state, municipal, and local organizations
	8. Non-governmental organizations	15. Number of international non-governmental organizations
		16. Number of national non-governmental organizations
		17. Number of state, municipal, and local non-governmental organizations
	9. Network structure	18. Links with financial institutions
		19. Links with raw material suppliers
		20. Links with technology providers
		21. Links with local, state, national, and international customers
		22. Links with other communities for knowledge exchange and transfer
	10. Property regime	23. Private property (ha)
		24. Ejido property (ha)
		25. Communal property (ha)
	11. Forest management operation rules	26. Forest cultivation rules
		27. Forest extraction rules
		28. Forest transformation rules
	12. Collective rules	29. Collective rules for conservation, management, and forest extraction in the communal statute
		30. Collective rules for conservation, management, and forest extraction in territorial planning
	13. Constitutional rules	31. Collective rules for conservation, management, and extraction in the National Forest Program
	14. Monitoring and sanction rules	32. Forest monitoring rules in the communal statute
		33. Rules for social, cultural, economic, and environmental organization in the communal statute
		34. Rules for sanctions and punishments of community members in the communal statute

(continued)

Table 12.2 (continued)

Category	Dimension	Indicator
6. Actors (A)	15. Number of relevant actors	35. Number of community members
		36. Number of ejido members
		37. Number of participating communities
	16. Past experiences	38. Number of years dedicated to cultivation, extraction, and transformation of wood
		39. Number of years in forest management
		40. Number of years in forest certification
	17. Collective leadership	41. Autonomous community organizational structure, self-managed and self-organized
	18. Dependence on timber forest resources	42. Income from timber forest resources (%)
7. Social, economic, and political environment (SEPE)	19. Economic development	43. Area of secondary trees (ha)
	20. Markets	44. Value of timber production ($)
	21. Technology	45. Use of advanced technology for cultivation, extraction, and transformation of wood (%)
		46. Use of basic technology for cultivation, extraction, and transformation of wood (%)
8. Related ecosystems (RE)	22. Climate patterns	47. Temperature (° C)
		48. Precipitation (mm)

Source Prepared by the authors based on McGinnis and Ostrom (2014)

through a drying process and transformation into *finger-joint* boards for furniture making.

2. **R.** *2. Forest sustainability.* The roundwood comes from their forests under forest management with the clearcutting method and includes pine species *Patula, Seudodtrobus, Ayacahuite, and Oaxacana,* the latter being the most extracted species. In 2000, the enterprise received the national forestry merit award from the Ministry of the Environment and Natural Resources (SEMARNAT by its acronym in Spanish) for sustainable forest management. Years later, it obtained the certification for good forest management from the Forest Stewardship Council through Rain Forest and Smartwood in 2001 and 2006. It also obtained the chain of custody certification from the same institution in 2006 and 2011.
3. **FRS.** *3. Limits of forest resource extraction, 4. Size of the forest resource system.* The enterprise belongs to the primary forestry sector, with a forest resource system size of 19,280 ha, of which 6,890.64 ha are montane cloud forest and jungles, and 12,389.50 ha are tropical forests (Technical Services 2010). According to the forest management program in 2006, 3,492.25 ha were allocated for forest extraction, with an extraction limit of up to a third of 6,832 ha.
4. **FRU.** *5. Growth of tree species and harvested pine volume* (m^3). The average annual harvested pine volume mainly ranges between 20,000 and 22,000 m^3 of log-board-tree. *6. Economic values.* The enterprise has an industrial area of 9,000 m^2, technological assets valued at 50.7 million of Mexican pesos, and a production capacity of 9,500 pieces of furniture per month in its three factories. The enterprise has made significant contracts with the Oaxaca State Public Education Institute; in 2006 it had a contract for 1.5 million dollars for the manufacture of school furniture.
5. **FGS.** *7. Governmental organizations, 8. Non-governmental organizations, 9. Network struc-*

ture, 10. Property regime, 11. Forest management operation rules, 12. Collective rules, 13. Constitutional rules, 14. Sanction and monitoring rules. The community's sustainable timber production is largely due to the forest governance system, which consists of a normative system of customs and a business system that organizes production under a communal land tenure regime. The forest governance system presents a series of information and knowledge transfer links to the communities and enterprise by federal institutions (CONABIO, SEDESOL, SEMARNAT, COINBIO, CONAFOR all of them, by their acronyms in Spanish), international non-governmental organizations (USAID, WWF, Forest Stewardship Council, Smartwood), municipal governmental institutions, local and communal authorities. The municipality of Ixtlán de Juárez has a communal statute in force since 2005, where the highest authority is the General Assembly of Commoners (GAC), composed of the Communal Property Commissariat (CPC), Surveillance Council (SC), and Communal Council (CC), positions that are renewed every three years, and the GAC is responsible for making decisions about forest management, establishing operational rules for cultivation, extraction, and transformation of wood, and collective rules for conservation, management, and forest extraction, within the framework of current environmental laws. The SC coordinates the forest rangers and technicians who are in charge of monitoring the forest resources; in addition to overseeing the actions of the commoners, they establish sanctions and punishments for commoners who do not participate in the activities inherent to forest management, which can range from the collection of monetary fines to expulsion from the community.

6. **A.** *15. Number of relevant actors, 16. Past experiences, 17. Collective leadership, 18. Dependence on timber forest resources.* The municipality has 4,500 inhabitants, of which 372 are commoners, of these 70 commoners hold positions in the GAC related to forest management, with leadership and experience of more than 25 years in silviculture. In the enterprise, 117 commoners work who are also partners of the enterprise, of which 44 are women and 73 are men.
7. **SEPE**. *19. Economic development, 20. Markets, 21. Technology.* The enterprise operates in market niches of medium and high-income customers who demand certified wood from sustainably managed forests. The enterprise's main markets are in the states of Oaxaca, Puebla, Guanajuato, and Mexico City. The sawmill has state-of-the-art technological equipment from Germany, and the wood-drying facility has stoves of Italian origin.
8. **RE**. *22. Climate patterns.* In Ixtlán, four types of climates are present that favor forest ecosystems and biodiversity: (i) intermediate humidity subhumid-temperate (oak-pine forest, pine-oak); (ii) high humidity subhumid-temperate (oak-pine forest, pine-oak forest, pine-fir); (iii) humid temperate (pine-oak forest, cloud forest); and (iv) warm humid (high evergreen jungle) (Aquino et al. 2012).

Therefore, the UCFAS is primarily dedicated to tree cultivation, logging, sawing, and furniture manufacturing for sale in the local market. The wood comes from forests under forest management programs that have delimited extraction and forestry use areas, such as the control of the annual harvested volume of roundwood. Ixtlán de Juárez has a communal statute in which the GAC establishes the operational rules, collective in congruence with the constitutional ones to which the commoners are subject to carry out forest management in accordance with forest development programs, and current environmental laws. The community has developed a system of forest governance of uses and customs, and business with a large number of links and collaboration networks with institutions at their different spatial scales for the generation and transfer of silvicultural knowledge. A third of the commoners with the most experience in silviculture work in the enterprise, with high participation of women. The growing demand for sawn wood and furniture has been met by good forestry practices and favorable climatic conditions, which has led to the expansion of their market niches at the national level, without undermining the sustainability of the forests.

12.5.2 UAFyDI

12.5.2.1 Emergence of the UAFyDI

The origin of the P'urhépecha community of San Juan Parangaricutiro dates back to the colonial era. Following the eruption of the Paricutín volcano in 1943, the community relocated to a place called Nuevo San Juan Parangaricutiro (NSJP), currently the head of the Parangaricutiro municipality (CONAFOR 2022). Legally recognized by presidential resolution in 1991, in accordance with the Agrarian Law (Chamber of Deputies 2023). In this sense, NSJP maintains the inalienability, imprescriptibility, and unseizability of communal lands, and the CPC as the representative and administrative management body of the GAC (UNDP 2012, 2023).

In 1981, the indigenous community of NSJP, while implementing the communal regime, began the implementation of the forestry enterprise with two separate economic units, the Forestry Use Unit and the Integral Development Unit. These were later integrated into a single entity known as the Forestry Use and Integral Development Unit (UAFyDI by its acronym in Spanish), responsible for activities throughout the timber production chain up to the obtaining of finished products (Anguiano-Cuara n.d.).

The community also has other non-timber forest product enterprises such as Mudemar and a resin and polymer distillery; agricultural extraction enterprises (orchards and fertilizer marketing), ecotourism, and service provision to the local population (television and internet, bottled water, and communal store). Other projects that could also be considered communal enterprises are related to urban transport service, agricultural improvement for the management and conservation of springs and the water cycle (Del-Castillo 2020), and sale of forest carbon (Climatereserve 2023).

12.5.2.2 Forest Sustainability

Focal action situations:

1. **I**. *1. Production, transformation, and sale of wood*. The communal enterprise produces wooden floors, pallets for the agri-food industry, moldings for the carpentry industry, and sawn wood for construction.
2. **R**. *2. Forest sustainability*. The roundwood comes from their forests under forest management certified by Forest Stewardship Council through Rain Forest and Smartwood. In 90% of the area, they apply the silvicultural development method, with a cutting cycle of 10 years. In the other 10% of the area, they apply the conservation method; where only 20% of the wood is extracted, as the objective is to maintain the natural and preserved forest. Notable pine species include *Pseudostrobus*, *Montezumae*, *leiophylla*, and *Michoacana*, in fir, A. *Religiosa*, and among the oaks, *Q. Candicans*, *Q. Rugosa*, and *Q. Laurina*; and among the Betulaceae, the *alder* (*Alnus jorullensis*) stands out.
3. **FRS**. *3. Limits of forest resource extraction*, *4. Size of the forest resource system*. The enterprise, belonging to the primary forest sector, has a forest resource system of 18,138 ha, of which 10,500 ha are pine surfaces (Fir, Oak, Quercus, and Alder). With the current forest management program in 2024, around 9,500 ha are under an intensive extraction method, another 1,000 ha are commercial forest plantations, and 382 ha are forests with high conservation value and are under strict protection by the community.
4. **FRU**. *5. Growth of tree species and harvested volume of pine (m^3)*. From its inception to the present, the average annual volume of authorized forest extraction is 65,000 m^3. *6. Economic values*. The enterprise has two sawmills, one with traditional technology and another with automated technology. It also has a sharpening workshop and a maintenance workshop, a spare-parts warehouse, a drying stove area, a secondary area, and a furniture factory.

5. **FGS**. *7. Governmental organizations, 8. Non-governmental organizations, 9. Network structure, 10. Property regime, 11. Operational rules of forest management, 12. Collective rules, 13. Constitutional rules, 14. Sanction and monitoring rules.* The community maintains a system of family possession in a communal regime dating back to 1981. The forest has been common property since 1991. Forest governance is made up of an internal government system of customs and traditions, and a business system. In the former, the highest decision-making body is the GAC. The board appointed by GAC is made up of the CPC, SC, and CC, whose board is renewed every three years and is responsible for making decisions about the community and with respect to the enterprise. In the latter, the enterprise's management is subordinate to the GAC and reports to the CPC and the SC. The communal statute in force since 2006 specifies the functional structure and the organs of enterprise administration. For forest management, the economic unit maintains a collaboration network with federal organizations (CONAFOR, INAES—all them by their acronym in Spanish), non-governmental organizations (Forest Stewardship Council, WWF), educational and research institutions (Autonomous University of Mexico, Autonomous University of Chapingo; Michoacana University of San Nicolás de Hidalgo), and local organizations (Alliance of Ejidos and Certified Forest Communities of Mexico). The communal statute of the community indicates that, based on the territorial planning study, customs and traditions and current legislation (Agrarian Law, General Law of Sustainable Forest Development and Law of the Social and Solidarity Economy); both the operational rules for the cultivation, extraction and transformation of the wood, and collective rules of conservation, management and forest extraction are established by the GAC, who must consider the needs of the commoners, communal enterprises, agreements signed with external organizations and other actors involved in forest management. Also, in said statute it is established that the SC is responsible not only for monitoring the actions of the enterprise's member commoners but also for forest monitoring.
6. **A**. *15. Number of relevant actors, 16. Past experiences, 17. Collective leadership, 18. Dependence on timber forest resources.* In 2020 NSJP had a total population of 16,745 inhabitants, of which approximately 7,500 were directly related to forest management, and 1,254 commoners were landowners of agricultural and forest lands. In this year, the UAFyDI provided 57% of the jobs generated by NSJP (approximately 680 people between permanent and temporary). Since the enterprise was founded to the present, for more than 50 years the commoners have accumulated knowledge about cultivation, extraction, transformation, and marketing of wood. Experiences and learnings from forest management that range from the struggle for the recognition of their communal lands, recognitions of merit and forest certification, to the drafting of the communal statute in 2006. Where, collective leadership has been fundamental to the achievement of their goals, based on the strengthening of indigenous identity, in which belonging to the territory, trust, reciprocity, and cooperation have played a fundamental role in the strengthening of the collective action of the NSJP commoners.
7. **SEPE**. *19. Economic development, 20. Markets, 21. Technology.* The main timber products such as wooden floors, pallets, sawn wood, and wooden moldings are mainly sold in Michoacán and Mexico City. However, the demand for their timber products is growing in the markets of the United States, Belgium, and Spain. For this, the enterprise has an automated sawmill that optimizes production processes and uses solar energy. It also employs new digital technologies in the development of forest management plans to georeference the places of wood extraction.
8. **ER**. *22. Climate patterns.* The forests of NSJP are at an average height of 2,550 masl under a temperate humid climate. Average temperature

of 18 °C and abundant rains in summer, which favor forest resources.

The UAFyDI is mainly dedicated to the cultivation, extraction of certified round and sawn wood from forests under forest management. It also transforms the wood for the marketing of wooden floors, pallets, and moldings. They have around 10,500 ha of pines, of which 9,500 ha are under intensive extraction, with an average annual volume of 65,000 m^3 of round wood. The forest governance system is made up of an internal government system of customs and traditions, and a business system, where the GAC is the main control and decision-making body. Decisions are made in accordance with its communal statute of 2006. It also establishes the operational and collective rules for forest monitoring. The enterprise, with over 50 years of experience in community forestry and around 20 years in forest management, has managed to develop a wide network of collaborations with local to international institutions. In 2020, the enterprise had 1,254 community members, many of whom have many years of experience in forest management, production, transformation, and marketing of timber products. The use of digital technologies for georeferencing timber extraction areas and automated machinery for its transformation has contributed to meeting national and foreign demand. Finally, favorable weather conditions have aided in the conservation and utilization of forests.

12.6 Discussions

12.6.1 Forest Sustainability of CENIE

The study analyzes the forest sustainability of CENIE based on the production of timber from forests certified under the SSEs framework (McGinnis and Ostrom 2014), which studies the relationship between the social and ecological elements of a community under communal land tenure. The aim is to examine the sustainability of forest resources based on the presence, relationships, and interactions between these two system elements. This analysis primarily differs from conventional sustainability assessments of food agroecosystems such as the MESMIS (by its acronym in Spanish) framework (Masera et al. 1999), comprehensive tools for evaluating the multidimensional performance of agroecosystems in terms of sustainability (Mottet et al. 2020), or the sustainability assessment of traditional agroecosystems (Palestina-González et al. 2021).

The study results show that, in the two cases of the enterprises studied, forest sustainability is related on one hand to the ecological systems present in the community, such as the large expanses of common-use forests that provide humans with ecosystem services of provision (wood, water, etc.), regulation (carbon capture, climate regulation, etc.), cultural (recreational), and support (nutrient recycling, soil formation, etc.) (Kandziora et al. 2013). It should be noted that, under pressure from peasant organizations, the Mexican state has played a fundamental role in creating the regulatory legal framework through land tenure reforms in 1992, which gave peasants certainty about the ownership of their communal lands (Bray et al. 2006). It placed the management of forest resources under their responsibility and opened the possibility of managing them for their conservation and utilization through the diversification of productive activities, strengthening, and creation of CENIE. On the other hand, forest sustainability is related to social systems. As mentioned earlier, following the land tenure reform, with the aim of extracting forest resources, communities strengthened their collective production organization system based on customs and traditions, which was recognized, for example, in the Political Constitution of the Free and Sovereign State of Oaxaca, in accordance with the principles of Article 115 of the Federal Constitution (Recondo 2001). Where the GAC is the institution responsible for establishing operational and collective rules in accordance with the constitutional ones for the management of forest resources. Over time, this social organization system of timber production evolved into a forest governance system, in which community members and other organizations related to forest conservation and extraction actively and collectively participate, creating and strengthening collaboration networks with different national and international organizations to incorporate new knowledge of forest management, sustainable technologies, and access to new market niches.

12.6.2 CENIE and the SDGs

Addressing the economic dimension of the SDGs, CENIE benefits communities by creating jobs and generating income for community members, contributing to the achievement of SDG 8 (decent work and economic growth) and SDG 1 (end of poverty). Regarding the social dimension, some enterprises have provided various services to the population (educational and technological), such as social infrastructure to communities, contributing to SDG 4 (quality education) and SDG 9 (industry, innovation, and infrastructure).

As for the ecological dimension, the enterprises have not only allowed the sustainable extraction of forest resources but have even increased the quantity and improved the quality of natural resources, contributing to SDG 13 (climate action) and SDG 15 (life on land). In response to criticisms of the SDGs, in the sense that they do not incorporate alternatives to development that are not based on conventional economic growth (considering the planet's limits), and that they also do not incorporate the cultural element (Hidalgo-Capitán et al. 2019). The results of this study show that promoting this type of communal business model is very convenient to reduce environmental problems, due to their contribution to the conservation and extraction of common-use natural resources. However, it is necessary to go beyond the SDGs. Indeed, due to the indigenous characteristic present in the two enterprises, which show a greater concern for nature care, they even let go of more profitable exploitation opportunities for the sustainability of forest resources. It should include the cultural component of indigenous peoples closely related to natural resources.

12.7 Conclusions

The purpose of the study was to examine the forest sustainability of two communal enterprises, UCFAS and UAFyDI located in Oaxaca and Michoacán, respectively. To achieve this, the SSEs framework was used, which aims to qualitatively evaluate the sustainability of common pool resources through the relationships and interactions found between the ecological and social systems of the system forestry. The two enterprises studied are engaged in the cultivation, extraction, transformation, and marketing of roundwood, saw timber, and finished products such as furniture, pallets, wooden floors, and moldings. They own large forest areas, averaging 18,000 ha. However, UCFAS allocates only 20% of its area to forestry, while UAFyDI allocates about 50% of its area to intensive forestry. UCFAS obtains an average annual harvested volume of pine wood of 20,000–22,000 m^3 of round-table-tree, and UAFyDI 65,000 m^3 of round-table-tree. This indicates that the Michoacan enterprise is highly extractive of its forest resources. However, in both cases, the wood comes from forests certified under forest management, in which extraction and forestry areas are delimited. The large extent of forests owned by the Michoacan enterprise may explain the higher extraction, another being the longevity and greater experience in cultivation and wood production. Given that, UAFyDI began operations in the first decade and UCFAS in the second decade of the eighties. Since then, the communities have been strengthening their forest governance systems, composed of the system of uses and customs, and the business system. Where the GAC is the highest consultative body and decision-making related to forestry and communal enterprises. Both communities had their communal statutes until the first decade of 2000 and are in force. From which, the GACs establish the operational and collective rules related to forest management and wood transformation. Both enterprises have around 50 years of experience in silviculture, and approximately 20 years in forest management. Over their trajectories, they have developed strong ties of collaboration with local to international institutions, for the transfer of knowledge and technologies.

Finally, in this study being a qualitative study, the analysis of the sustainability of common forest resources based on the 8 categories proposed by McGinnis and Ostrom (2014) was limited to the description of the relationships and interactions between the categories, and convincing explanations of the forest sustainability of the communal enterprises are found. Therefore, the study concludes that the CENIE have advanced toward the forest sustainability of timber production mainly related to forest governance system. However, the study's methodology does not allow a quantitative evaluation of the forest sustainability of each of the enterprises and jointly. Therefore, for future studies, it is suggested to construct an index to evaluate the sustainability of common forest resources, as has been done in some studies in Mexico (Leslie et al. 2015).

References

Abrahams C, Peredo AM (1996) Social work with poor women and their children: relevance of a developmental perspective. J Appl Soc Sci 21(1):53–59

Agrawal A (2001) Common property institutions and sustainable governance of resources. World Dev 29(10):1649–1672. https://doi.org/10.1016/s0305-750x(01)00063-8

Agrarian Law (2002) Anaya Publishers, Mexico City

Alston LJ (2018) New institutional economics. In: The new Palgrave dictionary of economics. Palgrave Macmillan. https://doi.org/10.1057/978-1-349-95189-5_2468

Anderies JM, Janssen MA, Ostrom E (2004) A framework to analyze the robustness of social-ecological systems from an institutional perspective. Ecol Soc 9(1):18. http://www.ecologyandsociety.org/vol9/iss1/art18

Anguiano-Cuara HA (n.d.) Indigenous Community Nuevo San Juan Parangaricutiro. Retrieved 20 Jan 2024 from https://www.gob.mx/cms/uploads/attachment/file/115013/Comunidad-H_ctor_Andr_s_Anguiano_Cuara.pdf

Antinori C, Bray D (2005) Community forest enterprises as entrepreneurial firms: economic and institutional perspectives from Mexico. World Dev 33(9):1529–1543. https://doi.org/10.1016/j.worlddev.2004.10.011

Aquino VC, Ruiz AF, Fuente CME (2012) Caracterización del patrimonio natural de la comunidad de Ixtlán de Juárez: Una aproximación desde el espacio territorial. In: Fuente CME, Ruiz AF, Aquino VC (eds) Conocimiento indígena contemporáneo y patrimonio biocultural en la Sierra Juárez de Oaxaca. Universidad de la Sierra Juárez, Oaxaca, México, pp 35–59

Azqueta D, Alviar M, Domínguez L, O'Ryan R (2007) Introduction to environmental economics. McGraw-Hill/Interamericana de España

Barkin D, Fuente M, Rosas M (2009) Tradition and innovation. Peasant contributions in guiding technological innovation to forge sustainability. Trajectories 11(299):39–54

Bassi I, Carestiato N (2016) Common property organisations as actors in rural development: a case study of a mountain area in Italy. Int J Commons 10(1):363–386. https://doi.org/10.18352/ijc.608

Berkes F, Folke C (1998) Linking social and ecological systems. Cambridge University Press, Cambridge, UK, Management practices and social mechanisms for building resilience

Bray DB, Antinori C, Torres-Rojo JM (2006) The Mexican model of community forest management: the role of agrarian policy, forest policy and entrepreneurial organization. Forest Policy Econ 8(4):470–484. https://doi.org/10.1016/j.forpol.2005.08.002

Caballero G, Garza MD (2010) The foundations of the new institutional economics towards the economics of natural resources: commons, institutions, governance and institutional change. University of Vigo, pp 1–26. https://www.ucipfg.com/Repositorio/MAES/MAES-09/Unidades_academicas/Unidad_1/Lectura3_unidad1Caballero.pdf

Caulley DN, Dowdy I (1987) Evaluation case histories as a parallel to legal case histories: accumulating knowledge and experience in the evaluation profession. Eval Program Plann 10(4):359–372. https://doi.org/10.1016/0149-7189(87)90008-5

Chamber of Deputies of the H. Congress of the Union (2023) Agrarian law. Off J Fed Latest Reform. Available at: https://www.diputados.gob.mx/LeyesBiblio/pdf/LAgra.pdf

Climatereserve (2023) San Juan Parangaricutiro [Video]. YouTube. https://www.youtube.com/watch?v=vMMJtvTwN3w

Cubbage FW, Davis RR, Paredes DR, Mollenhauer R, Elsin K, Frey GE, Salas DNC (2015) Community forestry enterprises in Mexico: sustainability and competitiveness. J Sustain For 34(6–7):623–650. https://doi.org/10.1080/10549811.2015.1040514

Del-Castillo A (2020) The Purepecha forest oasis that emerged from the ashes of a volcano. Verdebandera Environ J https://verdebandera.mx/el-oasis-forestal-purepecha-que-surgio-de-las-cenizas-de-un-volcan/

Denzin NK, Lincoln S (2012) General introduction Qualitative research as a discipline and practice. In: Manual of qualitative research. The field of qualitative research, vol 1, p 370. https://ehu.on.worldcat.org/oclc/796014766

Engbring G, Hajjar R (2021) Mexican community forest enterprises as social firms: organizational differences and the factors that shape them. For Policy Econ 131(102557). https://doi.org/10.1016/j.forpol.2021.102557

Engbring G, Hajjar R (2022) Conflicts between core purposes: trade-offs associated with organizational shifts in Mexican community forest enterprises. World Dev 160(106078). https://doi.org/10.1016/j.worlddev.2022.106078

Flick U (2007) Introduction to qualitative research, 2nd edn. Ediciones Morata, Fundación Paideia Galiza

Fuente NMM, Marin A (2024) Changing trees, enduring forests: institutional bricolage, gradual change and community forestry among Yucatec Mayans in Mexico. Dev Chang 55(1):97–122. https://doi.org/10.1111/dech.12815

Garibay C (2007) The corporate dilemma of forest communalism. Desacatos 23:251–274

Gómez-Mendoza L, Vega-Peña E, Ramírez IM, Palacio-Prieto JL, Galicia L (2006) Projecting land-use change processes in the Sierra Norte of Oaxaca, Mexico. Appl Geogr 26(3–4):276–290. https://doi.org/10.1016/j.apgeog.2006.09.003

Guber R (2011) Ethnography: method, field and reflexivity, 1st edn. Siglo XXI

Hernández HRB, Regino MJ, Miguel VAE (2020) The rural social enterprise and its contribution to the conservation of amaranth as agri-food heritage, case study: Centéotl. A.C. Zimatlán, Oaxaca, Mexico. J Andalusian Stud (REA) 39:166–179. https://doi.org/10.12795/rea.2020.i39.09

Hidalgo-Capitán AL, García-Álvarez S, Cubillo-Guevara AP, Medina-Carranco N (2019) The objectives of good living. An alternative proposal to the sustainable development goals. Ibero-Am J Dev Stud 8(1):6–57. https://doi.org/10.26754/ojs_ried/ijds.354

Jurjonas M, Pérez LM, Robson J, Noble AET (2023) Intergenerational perceptions of the collective action challenges facing Mexican community forests. Land Use Policy 134:106913. https://doi.org/10.1016/j.landusepol.2023.106913

Kandziora M, Burkhard B, Mülle F (2013) Interactions of ecosystem properties, ecosystem integrity and ecosystem service indicators—a theoretical matrix exercise. Ecol Ind 28:54–78. https://doi.org/10.1016/j.ecolind.2012.09.006

Leslie H, Basurto X, Nenadovic M, Sievanen L, Cavanaugh K, Cota-Nieto J, Erisman B, Finkbeiner E, Hinojosa-Arango G, Moreno-Baez M, Nagavarapu S, Redy S, Sánchez-Rodríguez A, Siegel K, Ulibarria-Valenzuela J, Weaver A, Aburto-Oropeza O (2015) Operationalizing the social-ecological systems framework to assess sustainability. PNAS 112(9):5979–5984. https://doi.org/10.1073/pnas.1414640112

Loreggian F, Secco L, Pettenella D (2023) Organizational models in European forestry: an attempt at conceptualization and categorization. Forests 14(5):905. https://doi.org/10.3390/f14050905

Lueck D (1994) Common property as an egalitarian share contract. J Econ Behav Organ 25(1):93–108. https://doi.org/10.1016/0167-2681(94)90088-4

Madrid L, Núñez JM, Quiroz G, Rodríguez Y (2009) Social forestry property in Mexico. Environ Res 1(2):179–196

Masera O, Astier M, López-Ridaura S (1999) Sustainability and natural resource management. MESMIS Eval Framework. Mundi-Prensa-GIRA-UNAM, Mexico City

McGinnis MD, Ostrom E (2014) Social-ecological system framework: initial changes and continuing challenges. Ecol Soc 19(2):30. https://doi.org/10.5751/es-06387-190230

Merino-Pérez L, Segura-Warnholtz G (2005) Forest and conservation policies and their impacts on forest communities in Mexico. In: Bray DB, Merino-Pérez L, Barry D (eds) The community forests of Mexico: managing for sustainable landscapes. University of Texas Press, Austin, Texas, USA, pp 49–69

Merino-Pérez L (2018) Forest communities in Mexico. Ways of life, governance, and conservation. Mex J Sociol 8(4):909–940

Mottet A, Bicksler A, Lucantoni D, De Rosa F, Scherf B, Scopel E, Tittonell P (2020) Assessing transitions to sustainable agricultural and food systems: a Tool for Agroecology Performance Evaluation (TAPE). Front Sustain Food Syst 4:252. https://doi.org/10.3389/fsufs.2020.579154

Muñoz-Piña C, de Janvry A, Sadoulet E (2003) Recrafting rights over common property resources in Mexico. Econ Dev Cult Change 2(1):129–158

National Forestry Commission (CONAFOR) (2013) Achievements and prospects of forest development in Mexico 2007–2012. General coordination of production and productivity. http://www.conafor.gob.mx:8080/biblioteca/ver.aspx?articulo=489

National Forestry Commission (CONAFOR) (2022) New San Juan Parangaricutiro: a model to follow. https://www.gob.mx/conafor/articulos/nuevo-san-juan-parangaricutiro-un-modelo-a-seguir

Oberlack C, Walter PL, Schmerbeck J, Tiwari B (2015) Institutions for sustainable forest governance: robustness, equity, and cross-level interactions in Mawlyngbna, Meghalaya, India. Int J Commons 9(2). https://doi.org/10.18352/ijc.538

Ostrom E (1998) A behavioral approach to the rational choice theory of collective action. Am Polit Sci Rev 92(1):1–22. https://doi.org/10.2307/2585925

Ostrom E (2009) A general framework for analyzing sustainability of social-ecological systems. Science 325(July):419–422. https://doi.org/10.1126/science.1172133

Ostrom E, Hess C (2007) Private and common property rights. https://doi.org/10.2139/ssrn.1304699

Ostrom E (2000) The governance of common goods. The evolution of collective action institutions. UNAM. FCE

Palacios-Izcara SP (2014) Manual of qualitative research, 1st edn. Fontamara

Palestina-González MI, Carranza-Cerda I, López-Reyes L, Torres E, Silva-Gómez SE (2021) Sustainability assessment of traditional agroecosystems in the high region of Yaonáhuac, Puebla, Mexico. Environments 8(5):40. https://doi.org/10.3390/environments8050040

Peredo AM, Chrisman JJ (2006) Toward a theory of community-based enterprise. Acad Manage Rev

31(2):309–328. https://doi.org/10.5465/amr.2006.20208683
Peredo AM, McLean M (2006) Social entrepreneurship: a critical review of the concept. J World Bus 41(1):56–65. https://doi.org/10.1016/j.jwb.2005.10.007
Recondo D (2001) Uses and customs, electoral processes and indigenous autonomy in Oaxaca, in Lourdes De León Pasquel, Customs, laws and Indian movement in Oaxaca and Chiapas, CIE-SAS-Miguel Ángel Porrúa, Mexico
Regino J, Wen Y, Cubbage F, Regino P (2017) Forest resources in the performance of Mexican community forest enterprises in a vertical integration system. Int J Sci 6(09):1–15. https://doi.org/10.18483/ijSci.1163
Regino J, Wen Y, Christopher C (2015) The role of forest resources in the performance of community forest enterprise in Mexico: analytical framework from competitive strategy. Int J Sci 4(05):14–24. https://doi.org/10.18483/ijSci.700
Segura-Warnholtz G (2014) Fifteen years of public policies for collective action in forest communities. Mex J Sociol 76(5):105–135
SEMARNAT (2003) Deforestation in 24 PRODERS Regions. National Commission of Natural Protected Areas, Mexico, 123 pp.
Technical Forestry Services (2010) Main ecosystems and land use by area in the community of Ixtlán de Juárez, Oaxaca
Tetreault D (2004) A taxonomy of sustainable development models. Spiral Stud State Soc X(29):45–77
Toledo V, Ortíz-Espejel B (2014) Mexico, regions that are moving towards sustainability. Iberoamerican University Puebla, a geopolitics of biocultural resistances
UNDP (2012) Indigenous community of Nuevo San Juan Parangaricutiro. https://www.equatorinitiative.org/wp-content/uploads/2017/05/case_1_1363201559.pdf
UNDP (2023) Municipal human development report 2010–2020: a decade of local transformations in Mexico. https://www.undp.org/es/mexico/publicaciones/informe-de-desarrollo-humano-municipal-2010-2020-una-decada-de-transformaciones-locales-en-mexico-0
Vega DEC, Keenan RJ (2016) Situating community forestry enterprises within new institutional economic theory: what are the implications for their organization? J For Econ 25:1–13. https://doi.org/10.1016/j.jfe.2016.07.001
White A, Martin A (2002) Who owns the world's forests? forest trends. Center for International Law, Washington, DC
Wilshusen PR (2003) Negotiating devolution: community conflict, structural power, and local forest management in Quintana Roo. University of Michigan, School of Natural Resources and Environment, Ann Arbor, Mexico
Yin R (n.d.) Case study research. Design and methods. SAGE publications. International Educational and Professional Publisher, Thousand Oaks, London, New Delhi. file:///C:/Users/juanr/Downloads/YIN%20ROBERT%20.pdf
Zhang Y, Uusivuori J, Kuuluvainen J (2000) Impacts of economic reforms on rural forestry in China. For Policy Econ 1(1):27–40. https://doi.org/10.1016/s1389-9341(00)00007-1
Zhang S, Paterson JS, Hujala T (2021) Sustaining forest ecosystem services through social enterprises: motivations and challenges from a case study in Scotland. Small-Scale For 20(4):627–647. https://doi.org/10.1007/s11842-021-09484-7

The manufacturer's authorised representative in the EU is Springer
Nature Customer Service Centre GmbH, Europaplatz 3, 69115 Heidelberg,
Germany. If you have any concerns regarding our products, please
contact ProductSafety@springernature.com

Printed and bound by CPI Group (UK) Ltd, Croydon, CR0 4YY
23/09/2026
02221277-0001